STUDENT'S SOLUTIONS MANUAL

PAMELA OMER

Western New England University

INTRODUCTORY STATISTICS:
EXPLORING THE WORD THROUGH DATA
THIRD EDITION

Robert Gould

University of California, Los Angeles

Rebecca Wong

West Valley College

Colleen Ryan

Moorpark Community College

 Pearson

The author and publisher of this book have used their best efforts in preparing this book. These efforts include the development, research, and testing of the theories and programs to determine their effectiveness. The author and publisher make no warranty of any kind, expressed or implied, with regard to these programs or the documentation contained in this book. The author and publisher shall not be liable in any event for incidental or consequential damages in connection with, or arising out of, the furnishing, performance, or use of these programs.

Reproduced by Pearson from electronic files supplied by the author.

ISBN-13: 978-0-13-518923-8
ISBN-10: 0-13-518923-3

CONTENTS

Chapter 1: Introduction to Data

Section 1.2: Classifying and Storing Data 1
Section 1.3: Investigating Data 2
Section 1.4: Organizing Categorical Data 2
Section 1.5: Collecting Data to Understand Causality 3
Chapter Review Exercises 4

Chapter 2: Picturing Variation with Graphs

Section 2.1: Visualizing Variation in Numerical Data
    and Section 2.2: Summarizing Important Features of a Numerical Distribution 5
Section 2.3: Visualizing Variation in Categorical Variables
    and Section 2.4: Summarizing Categorical Distributions.....................................8
Section 2.5: Interpreting Graphs 8
Chapter Review Exercises 8

Chapter 3: Numerical Summaries of Center and Variation

Section 3.1: Summaries for Symmetric Distributions 11
Section 3.2: What's Unusual? The Empirical Rule and z-Scores...............................12
Section 3.3: Summaries for Skewed Distributions 13
Section 3.4: Comparing Measures of Center 13
Section 3.5: Using Boxplots for Displaying Summaries 15
Chapter Review Exercises 15

Chapter 4: Regression Analysis: Exploring Associations
            between Variables

Section 4.1: Visualizing Variability with a Scatterplot 17
Section 4.2: Measuring Strength of Association with Correlation 17
Section 4.3: Modeling Linear Trends 17
Section 4.4: Evaluating the Linear Model 20
Chapter Review Exercises 22

Chapter 5: Modeling Variation with Probability

Section 5.1: What Is Randomness?..27
Section 5.2: Finding Theoretical Probabilities..27
Section 5.3: Associations in Categorical Variables 30
Section 5.4: Finding Empirical and Simulated Probabilities....................................31
Chapter Review Exercises 32

Chapter 6: Modeling Random Events: The Normal and Binomial Models

Section 6.1: Probability Distributions Are Models of Random Experiments............35
Section 6.2: The Normal Model...36
Section 6.3: The Binomial Model (Optional) 43
Chapter Review Exercises 44

Chapter 7: Survey Sampling and Inference

Section 7.1: Learning about the World through Surveys.............................47
Section 7.2: Measuring the Quality of a Survey 48
Section 7.3: The Central Limit Theorem for Sample Proportions...............48
Section 7.4: Estimating the Population Proportion with Confidence Intervals 50
Section 7.5: Comparing Two Population Proportions with Confidence....................52
Chapter Review Exercises 53

Chapter 8: Hypothesis Testing for Population Proportions

Section 8.1: The Essential Ingredients of Hypothesis Testing 55
Section 8.2: Hypothesis Testing in Four Steps 56
Section 8.3: Hypothesis Tests in Detail 58
Section 8.4: Comparing Proportions from Two Populations.......................59
Chapter Review Exercises 61

Chapter 9: Inferring Population Means

Section 9.1: Sample Means of Random Samples 67
Section 9.2: The Central Limit Theorem for Sample Means........................67
Section 9.3: Answering Questions about the Mean of a Population.............68
Section 9.4: Hypothesis Testing for Means 69
Section 9.5: Comparing Two Population Means 72
Chapter Review Exercises 76

Chapter 10: Associations between Categorical Variables

Section 10.1: The Basic Ingredients for Testing with Categorical Variables..............81
Section 10.2: The Chi-Square Test for Goodness of Fit...............................82
Section 10.3: Chi-Square Tests for Associations between
    Categorical Variables...84
Section 10.4: Hypothesis Tests When Sample Sizes Are Small...................89
Chapter Review Exercises 91

Chapter 11: Multiple Comparisons and Analysis of Variance

Section 11.1: Multiple Comparisons..95
Section 11.2: The Analysis of Variance 96
Section 11.3: The ANOVA Test..96
Section 11.4: Post-Hoc Procedures..98
Chapter Review Exercises 100

Chapter 12: Experimental Design: Controlling Variation

Section 12.1: Variation Out of Control..101
Section 12.2: Controlling Variation in Surveys..103
Section 12.3: Reading Research Papers...103

Chapter 13: Inference without Normality

Section 13.1: Transforming Data...107
Section 13.2: The Sign Test for Paired Data...108
Section 13.3: Mann-Whitney Test for Two Independent Groups...........................109
Section 13.4: Randomization Tests...110
Chapter Review Exercises 110

Chapter 14: Inference for Regression

Section 14.1: The Linear Regression Model..115
Section 14.2: Using the Linear Model 115
Section 14.3: Predicting Values and Estimating Means 116
Chapter Review Exercises 117

Chapter 1: Introduction to Data

Section 1.2: Classifying and Storing Data

1.1 There are eight variables: "Female", "Commute Distance", "Hair Color", "Ring Size", "Height", "Number of Aunts", "College Units Acquired", and "Living Situation".

1.3 a. Living situation is categorical.

   b. Commute distance is numerical.

   c. Number of aunts is numerical.

1.5 Answers will vary but could include such things as number of friends on Facebook or foot length. *Don't copy these answers*.

1.7 0 = male, 1 = female. The sum represents the total number females in the data set.

1.9 Female is categorical with two categories. The 1's represent females, and the 0's represent males. If you added the numbers, you would get the number of females, so it makes sense here.

1.11 a. The data is stacked.

   b. 1 = male, 0 = female.

   c.

Male	Female
1916	9802
183	153
836	1221
95	
512	

1.13 a. Stacked and coded:

Calories	Sweet
90	1
310	1
500	1
500	1
600	1
90	1
150	0
600	0
500	0
550	0

The second column could be labeled "Salty" with the 1's being 0's and the 0's being 1's.

   b. Unstacked:

Sweet	Salty
90	150
310	600
500	500
500	550
600	
90	

Section 1.3: Investigating Data

1.15 Yes. Use College Units Acquired and Living Situation.

1.17 No. Data on number of hours of study per week are not included in the table.

1.19 a.   Yes. Use Date.

b.   No. data on temperature are not included in the table.

c.   Yes. Use Fatal and Species of Shark.

d.   Yes. Use Location.

Section 1.4: Organizing Categorical Data

1.21 a.   33/40 = 82.5%

b.   32/45 = 71.1%

c.   33/65 = 50.8%

d.   82.5% of 250 = 206

1.23 a.   15/38 = 39.5% of the class were male.

b.   0.64(234) = 149.994, so 150 men are in the class.

c.   0.40(x) = 20, so 20/0.40 = 50 total students in the class.

1.25 The frequency of women 6, the proportion is 6/11, and the percentage is 54.5%.

1.27 a. and b.

	Men	Women	Total
Dorm	3	4	7
Commuter	2	2	4
Total	5	6	11

c.   4/6 = 66.7%

d.   4/7 = 57.1%

e.   7/11 = 63.6%

f.   66.7% of 70 = 47

1.29   1.26(x) = 160328 so 160328/1.26 = 127,244 personal care aids in 2014

1.31

State	Prison	Rank Prison	Population	Population (thousands)	Prison per 1000	Rank Rate
California	136,088	1	39,144,818	39145	3.48	4
New York	52518	2	19,795,791	19796	2.65	5
Illinois	48278	3	12,859,995	12860	3.75	3
Louisiana	30030	4	4,670,724	4671	6.43	1
Mississippi	18793	5	2,992,333	29922	6.28	2

California has the highest prison population. Louisiana has the highest rate of imprisonment.

The two answers are different because the state populations are different.

1.33

Year	%Uncovered
1990	$\dfrac{34,719}{249,778} = 13.9\%$
2000	$\dfrac{36,586}{279,282} = 13.1\%$
2015	$\dfrac{29758}{316574} = 9.4\%$

The percentage of uninsured people have been declining.

1.35

Year	%Older Population
2020	$\dfrac{54.8}{334} = 16.4\%$
2030	$\dfrac{70.0}{358} = 19.6\%$
2040	$\dfrac{81.2}{380} = 21.4\%$
2050	$\dfrac{88.5}{400} = 22.1\%$

The percentage of older population is projected to increase.

1.37   We don't know the percentage of female students in the two classes. The larger number of women at 8a.m. may just result from a larger number of students at 8 a.m., which may be because the class can accommodate more students because perhaps it is in a large lecture hall.

Section 1.5 Collecting Data to Understand Causality

1.39   Observational study.

1.41   Controlled experiment.

1.43   Controlled experiment.

1.45   Anecdotal evidence are stories about individual cases. No cause-and effect conclusions can be drawn from anecdotal evidence.

1.47   This was an observational study, and from it you cannot conclude that the tutoring raises the grades. Possible confounders (answers may vary): 1. It may be the more highly motivated who attend the tutoring, and this motivation is what causes the grades to go up. 2. It could be that those with more time attend the tutoring, and it is the increased time studying that causes the grades to go up.

1.49   a.   The sample size of this study is not large (40). The study was a controlled experiment and used random assignment. It was not double-blind since researchers new what group each participant was in.

     b.   The sample size of the study was small, so we should not conclude that physical activity while learning caused higher performance.

1.51   a.   Controlled experiment. Researchers used random assignment of subjects to treatment or control groups.

     b.   Yes. The experiment had a large sample size, was controlled, randomized, and double-blind; and used a placebo.

1.53   No, this was not a controlled experiment. There was no random assignment to treatment/control groups and no use of a placebo.

1.55 a.   Intervention remission: $11/33 = 33.3\%$; Control remission: $3/34 = 8.8\%$

      b.   Controlled experiment. There was random assignment to treatment/control groups.

      c.   While this study did use random assignment to treatment/control groups, the sample size was fairly small (67 total) and there was no blinding in the experimental design. The difference in remission may indicate that the diet approach is promising and further research in this area is needed.

1.57 No. This is an observational study.

Chapter Review Exercises

1.59 a. $61/98 = 62.2\%$

      b. $37/82 = 45.1\%$

      c.   Yes, this was a controlled experiment with random assignment. The difference in percentage of homes adopting smoking restrictions indicates the intervention may have been effective.

1.61 a.   Gender (categorical) and whether students had received a speeding ticket (categorical)

      b.

	Male	Female
Yes	6	5
No	4	10

      c.   Men: $6/10=60\%$; Women: $5/15 = 33.3\%$; a greater percentage of men reported receiving a speeding ticket.

1.63 Answers will vary. *Students should not copy the words they see in these answers.* Randomly divide the group in half, using a coin flip for each woman: Heads she gets the vitamin D, and tails she gets the placebo (or vice versa). Make sure that neither the women themselves nor any of the people who come in contact with them know whether they got the treatment or the placebo ("double-blind"). Over a given length of time (such as three years), note which women had broken bones and which did not. Compare the percentage of women with broken bones in the vitamin D group with the percentage of women with broken bones in the placebo group.

1.65 a.   The treatment variable is mindful yoga participation. The response variable is alcohol use.

      b.   Controlled experiment (random assignment to treatment/control groups).

      c.   No, since the sample size was fairly small; however, the difference in outcomes for treatment/control groups may indicate that further research into the use of mindful yoga may be warranted.

1.67 No. There was no control group and no random assignment to treatment or control groups.

1.69 a.   LD: 8% tumors; LL: 28% tumors A greater percentage of the 24 hours of light developed tumors.

      b.   A controlled experiment. You can tell by the random assignment.

      c.   Yes, we can conclude cause and effect because it was a controlled experiment, and random assignment will balance out potential confounding variables.

Chapter 2: Picturing Variation with Graphs

Section 2.1: Visualizing Variation in Numerical Data
and Section 2.2: Summarizing Important Features of a Numerical Distribution

2.1 a. 4 people had resting pulse rates more than 100.

  b. $\dfrac{4}{125} = 3.2\%$ of the people had resting pulse rates of more than 100.

2.3 New vertical axis labels: $\dfrac{1}{25} = 0.04$, $\dfrac{2}{25} = 0.08$, $\dfrac{3}{25} = 0.12$, $\dfrac{4}{25} = 0.16$, $\dfrac{5}{25} = 0.20$

2.5 Yes, since only about 7% of the pulse rates were higher than 90 bpm. Conclusion might vary, but students must mention that 7% of pulse rates were higher than 90 bpm.

2.7 a. Both cereals have similar center values (about 110 calories). The spread of the dotplots differ.

  b. Cereal from manufacturer K tend to have more variation.

2.9 Roughly bell shaped. The lower bound is 0, the mean will be a number probably below 9, but a few students might have slept quite a bit (up to 12 hours?) which creates a right-skew.

2.11 It would be bimodal because the men and women tend to have different heights and therefore different arm spans.

2.13 About 75 beats per minute.

2.15 The BMI for both groups are right skewed. For the men it is maybe bimodal (hard to tell). The typical values for the men and women are similar although the value for the men appears just a little bit larger than the typical value for the women. The women's values are more spread out.

2.17 a. The distribution is multimodal with modes at 12 years (high school), 14 years (junior college), 16 years (bachelor's degree), and 18 years (possible master's degree). It is also left-skewed with numbers as low as 0.

  b. Estimate: 300 + 50 + 100 + 40 + 50, or about 500 to 600, had 16 or more years.

  c. Between $\dfrac{500}{2018}$, or about 25%, and $\dfrac{600}{2018}$, or about 30%, have a bachelor's degree or higher. This is very similar to the 27% given.

2.19 Ford typically has higher monthly costs (the center is near 250 dollars compared with 225 for BMW) and more variation in monthly costs.

2.21 1. The assessed values of homes would tend to be lower with a few higher values: This is histogram B.

  2. The number of bedrooms in the houses would be slightly skewed right: This is histogram A.

  3. The height of house (in stories) for a region would be that allows up to 3 stories would be histogram C.

2.23 1. The heights of students would be bimodal and roughly symmetrical: This is histogram B.

  2. The number of hours of sleep would be unimodal and roughly symmetrical, with any outliers more likely being fewer hours of sleep: This is histogram A.

  3. The number of accidents would be left skewed, with most student being involved in no or a few accidents: This is histogram C.

2.25   Students should display a pair of dotplots or histograms. One graph for Hockey and one for Soccer. The hockey team tends to be heavier than the soccer team (the typical hockey player weighs about 202 pounds while the typical soccer player weighs about 170 pounds). The soccer team has more variation in weights than the hockey team because there is more horizontal spread in the data. Statistical Question (answers may vary): Are hockey players heavier than soccer players? Which type of athlete has the most variability in weight?

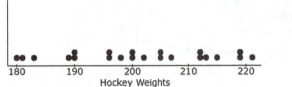
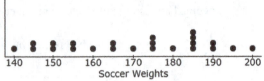

2.27   See histogram. The shape will depend on the binning used. The histogram is bimodal with modes at about $30 and about $90.

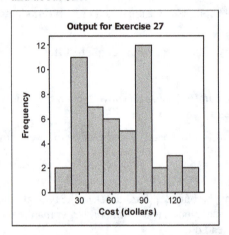

2.29   See histogram. The histogram is right-skewed. The typical value is around 12 (between 10 and 15) years, and there are three outliers: Asian elephant (40 years), African elephant (35 years), and hippo (41 years). Humans (75 years) would be way off to the right; they live much longer than other mammals.

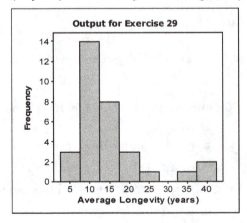

2.31   Both graphs are multimodal and right-skewed. The Democrats have a higher typical value, as shown by the fact that the center is roughly around 35 or 40%, while the center value for the Republicans is closer to 20 to 30%. Also note the much larger proportion of Democrats who think the rate should be 50% or higher. The distribution for the Democrats appears more spread out because the Democrats have more people responding with both lower and higher percentages.

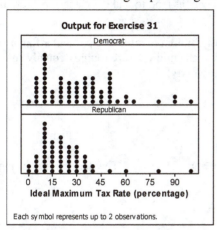

2.33   The distribution appears left-skewed because of the low-end outlier at about $20,000 (Brigham Young University).

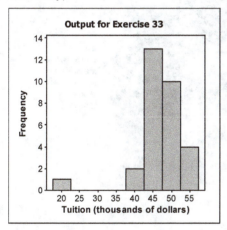

2.35   With this grouping the distribution appears bimodal with modes at about 110 and 150 calories. (With fewer—that is, wider—bins, it may not appear bimodal.) There is a low-end outlier at about 70 calories. There is a bit of left skew.

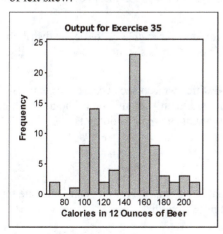

Section 2.3: Visualizing Variation in Categorical Variables and Section 2.4: Summarizing Categorical Distributions

2.37   No, the largest category is Wrong to Right, which suggests that changes tend to make the answers more likely to be right.

2.39   A bar graph with the least variability would be one in which most children favored one particular flavor (like chocolate). A bar graph with most variability, would be one in which children were roughly equally split in their preference. with 1/3 choosing vanilla, 1/3 chocolate, 1/3 strawberry.

2.41   a.   "All of the Time" was the most frequent response.

   b.   The difference is about 10%. It is easier to see in the bar chart.

2.43   a.   40–59-year old's.

   b.   The obesity rates for women are slightly higher in the 20–39 and 60+ groups. The obesity rate for men is higher in the 40–59 age group.

2.45   Bar graph or pie chart. Chrome controls the highest market share.

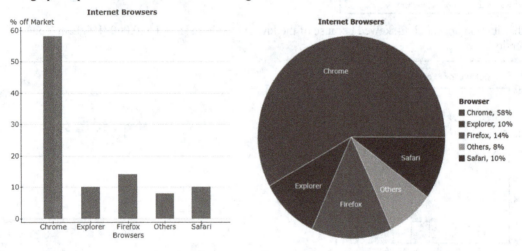

Section 2.5: Interpreting Graphs

2.47   This is a histogram, which we can see because the bars touch. The software treated the values of the variable *Garage* as numbers. However, we wish them to be seen as categories. A bar graph or pie chart would be better for displaying the distribution.

2.49   Hours of sleep is a numerical variable. A histogram or dotplot would better enable us to see the distribution of values. Because there are so many possible numerical values, this pie chart has so many "slices" that it is difficult to tell which is which.

2.51   Those who still play tended to have practiced more as teenagers, which we can see because the center of the distribution for those who still play is about 2 or 2.5 hours, compared to only about 1 or 1.5 hours for those who do not. The distribution could be displayed as a pair of histograms or a pair of dotplots.

Chapter Review Exercises

2.53   Since the data are numerical, a pair of histograms or dotplots could be used, one for the males and one for the females. A statistical question is Who slept more on average, men or women?"

2.55   a.   The diseases with higher rates for HRT were heart disease, stroke, pulmonary embolism, and breast cancer. The diseases with lower rates for HRT were endometrial cancer, colorectal cancer, and hip fracture.

     b.   Comparing the rates makes more sense than comparing just the numbers, in case there were more women in one group than in the other.

2.57   The vertical axis does not start at zero and exaggerates the differences. Make a graph for which the vertical axis starts at zero.

2.59   The shapes are roughly bell-shaped and symmetric; the later period is warmer, but the spread is similar. This is consistent with theories on global warming. The difference is 57.9 – 56.7 = 1.2, so the difference is only a bit more than 1degree Fahrenheit.

2.61   a.   A smaller percentage favor nuclear energy in 2016.

     b.   The Republicans show the most change.

     c.   A side-by-side bar graph (Republican 2010 adjacent to Republican 2016) would make the comparison easier.

2.63   The created 10-point dotplot will vary, but the dotplot for this exercise should be right-skewed.

2.65   Graphs will vary. Histograms and dotplots are both appropriate. For the group without a camera the distribution is roughly symmetrical, and for the group with a camera it is right-skewed. Both are unimodal. The number of cars going through a yellow light tends to be less at intersections with cameras. Also, there is more variation in the intersections without cameras.

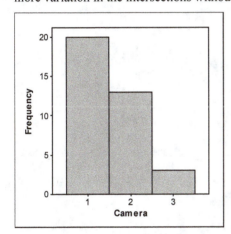 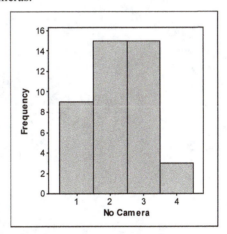

2.67   Both distributions are right-skewed. The typical speed for the men (a little above 100 mph) is a bit higher than the typical speed for the women (which appears to be closer to 90 mph). The spread for the men is larger primarily because of the outlier of 200 mph for the men.

2.69   The distribution should be right-skewed.

2.71   a.   The tallest bar is Wrong to Right, which suggests that the instruction was correct.

     b.   For both instructors, the largest group is Wrong to Right, so it appears that changes made tend to raise the grades of the students.

2.73   a.   Facebook (only about 5% used it less often than weekly).

     b.   LinkedIn (only about 20% used it daily).

     c.   Facebook (around 75% were in one category—daily).

2.75   a.   Histogram or dotplot

     b.   Side-by-side barplots showing gender frequencies separately for each zip code.

Chapter 3: Numerical Summaries of Center and Variation

Answers may vary slightly, especially for quartiles and interquartile ranges, due to type of technology used, or rounding.

Section 3.1: Summaries for Symmetric Distributions

3.1   c

3.3   The mean is between about 4 and 6 hours.

3.5   a.   The mean number of floors is 118.6.

   b.   The standard deviation of the number of floors is 26.0.

   c.   Dubai is farthest from the mean.

Column	n	Mean	Variance	Std. dev.	Std. err.	Median	Range	Min	Max	Q1	Q3
# Floors	5	118.6	676.3	26.005769	11.630133	108	62	101	163	101	120

3.7   a.   The typical river is 2230.8 miles long.

   b.   The standard deviation is 957.4 miles. The Mississippi-Missouri-Red Rock River contributes most to the standard deviation because it is farthest from the mean.

Column	n	Mean	Variance	Std. dev.	Std. err.	Median	Range	Min	Max	Q1	Q3
Length (in miles)	5	2230.8	916540.7	957.36132	428.145	1900	2260	1450	3710	1459	2635

   c.   The mean would decrease, and the standard deviation will increase. New mean: 1992.3 miles, new standard deviation: 1036.5 miles.

Column	n	Mean	Variance	Std. dev.	Std. err.	Median	Range	Min	Max	Q1	Q3
Length (in miles)	5	2230.8	916540.7	957.36132	428.145	1900	2260	1450	3710	1459	2635

3.9   a.   For the early 1900s the mean is 22.02 seconds; the standard deviation is 0.40 seconds.

Column	n	Mean	Variance	Std. dev.	Std. err.	Median	Range	Min	Max	Q1	Q3
Time (1900-1920)	5	22.02	0.162	0.40249224	0.18	22	1	21.6	22.6	21.7	22.2

   b.   For recent Olympics, the mean is 19.66 seconds; the standard deviation is 0.35 seconds.

Column	n	Mean	Variance	Std. dev.	Std. err.	Median	Range	Min	Max	Q1	Q3
Time (2000-2016)	5	19.66	0.123	0.35071356	0.15684387	19.8	0.8	19.3	20.1	19.3	19.8

   c.   Recent winners are faster and have less variation in their winning times.

3.11   a.   A total of 185 people were surveyed.

   b.   Comparing the means, men thought more should be spent on a wedding. Comparing the standard deviations, men had more variation in their responses.

3.13   a.   The mean for longboards is 12.4 days, which is more than the mean for shortboards, which is 9.8 days. So longboarders tend to get more surfing days.

   b.   The standard deviation of 5.2 days for the longboarders was larger than the standard deviation of 4.2 days for the shortboarders. So the longboarders have more variation in days.

```
Descriptive Statistics: Long, Short
Variable   N     Mean    StDev   Minimum     Q1    Median      Q3    Maximum
Long       30   12.367   5.249     4.000  8.000   11.500  16.750   22.000
Short      30    9.767   4.256     4.000  6.750    9.500  12.000   20.000
```

3.15   San Jose tends to have a higher typical temperature; Denver has more variation in temperature.

3.17   a.   20.3 to 40.1 pounds.

   b.   Less than one standard deviation from the mean.

3.19   a.   The mean is $3.66 and represents the typical price of a container (59 to 64 ounces) of orange juice at this site.

   b.   The standard deviation is 0.51. Most orange juice of this size is within 51 cents of $3.66.

Column	n	Mean	Variance	Std. dev.	Std. err.	Median	Range	Min	Max	Q1	Q3
OJ Prices ($)	10	3.659	0.26183222	0.51169544	0.16181231	3.785	1.5	2.99	4.49	2.99	3.99

3.21   The standard deviation for the 100-meter event would be less. All the runners come to the finish line within a few seconds of each other. In the marathon, the runners can be quite widely spread out after running that long distance.

3.23   South Carolina: Mean is $198.1 thousand; standard deviation is $68.0 thousand. Tennessee: Mean is $215.8 thousand; standard deviation is $75.1 thousand. Houses in Tennessee are typically more expensive and have more variation in price than houses in South Carolina.

Column ◆	n ◆	Mean ◆	Variance ◆	Std. dev. ◆	Std. err. ◆	Median ◆	Range ◆	Min ◆	Max ◆	Q1 ◆	Q3 ◆
SC (in $thousands)	15	198.13333	4625.5524	68.011414	17.560471	183	224	110	334	160	221
TN (in $thousands)	15	215.8	5640.0286	75.100124	19.390769	200	275	125	400	160	280

3.25   a.   The mean for the men was 10.5, and for the women it was 4.7, showing that the male drinkers typically drank more (on average, almost six drinks more) than the female drinkers.

   b.   The standard deviation for the men was 11.8, and the standard deviation for the women was 4.8, showing much more variation in the number of drinks for the men.

```
Descriptive Statistics: Drinks_Female, Drinks_Male
Variable          N    Mean    StDev
Drinks_Female     46   4.696   4.821
Drinks_Male       53   10.55   11.83
```

   c.   The mean for the men is now 8.6, and the mean for the women is still 4.7. Thus the mean for the men is still larger than the mean for the women, but not as much larger.

   d.   The standard deviation would be smaller without the two outliers. This is because the contribution to the standard deviation from these two outliers is large since they are farthest from the mean, and that contribution would be removed.

```
Descriptive Statistics: Drinks_Female, Drinks_Male w/o 48 & 70
Variable                    N    Mean    StDev
Drinks_Female               46   4.696   4.821
Drinks_Male w/o 48 & 70     51   8.647   6.57
```

3.27   A standard deviation can be zero if all the values are the same (no variation between the data values).

Section 3.2: What's Unusual? The Empirical Rule and z-Scores

3.29   a.   68%

   b. $(16 + 34 + 27)/120 = 64\%$. The estimate is very close to 68%.

   c.   Between 555 and 819 runs.

3.31   a.   Approximately 95%, using the empirical rule $(35.9 \pm 2(11.6))$

   b.   Approximately 68% using the empirical rule (35.9 ± 11.6)

c. No, because it is not more than 2 standard deviations from the mean.

3.33 a. $\dfrac{58-64}{3} = \dfrac{-6}{3} = -2$

b. $x = \bar{x} + zs = 64 + 1(3) = 64 + 3 = 67$ inches (or 5 feet 7 inches)

3.35 An IQ below 80 is more unusual because 80 is 1.33 standard deviations from the mean while 110 is only 0.67 standard deviations from the mean.
$$80 - 100 / 15 = -1.33 \qquad 110 - 100 / 15 = .67$$

3.37 a. $z = \dfrac{2500 - 3462}{500} = \dfrac{-962}{500} = -1.92$

b. $z = \dfrac{2500 - 2622}{500} = \dfrac{-122}{500} = -0.24$

c. A birth rate of 2500 grams is more common (the z-score is closer to 0) for babies born one month early. In other words, there is a higher percentage of babies with low birth weight among those born one month early. This makes sense because babies gain weight during gestation, and babies born one month early have had less time to gain weight.

3.39 a. $69 = 2(3) = 75$ inches

b. $69 - (1.5)3 = 64.5$ inches

Section 3.3: Summaries for Skewed Distributions

3.41 Two measures of the center of data are the mean and the median. The median is preferred for data that are strongly skewed or have outliers. If the data are relatively symmetric, the mean is preferred but the median is also okay.

3.43

363	384	389	408	423	434	471	520	602	677
		Q1		Med		Q3			

a. Median: 428.5 million; about half the top 10 Marvel movies made more than $428.5 million.

b. Q1 = 389 million, Q3 = 520 million; IQR = 520 – 389 = 131 million. This is the range of the middle 50% of the sorted incomes in the top 10 Marvel movies.

c. Range = 677 – 363 = 314 million. The IQR is preferred over the range because the range depends on only two observations and because it very sensitive to any extreme values in the data.

3.45 Median = 471 million. About 50% of the top 7 Marvel movies made more than $471 million.

3.47 a. 25%

b. 75%

c. 50%

d. IQR = Q3 – Q1 = 390 – 237.7 = 152.3. The range of the middle 50% of the sorted data is 152.3 million BTUs.

Section 3.4: Comparing Measures of Center

3.49 a. Outliers are observed values that are far from the main group of data. In a histogram they are separated from the others by space. If they are mistakes, they should be removed. If they are not mistakes do the analysis twice: once with and once without outliers.

   b.   The median is more resistant, which implies that it changes less than the mean (when the data with and without outliers are compared).

3.51 The corrected value will give a different mean but not a different median. Medians are not as affected by the size of extreme scores, but the mean is affected.

3.53 a.   The distribution is right-skewed.

   b.   The median and the IQR should be used to describe the distribution.

3.55 a.   The distributions are right-skewed.

   b.   The medians.

   c.   The interquartile ranges.

   d.   The typical Democratic senator has been in office 9 years, while the typical Republican senator has been in office 6 years. There is more variability in the experience of Democratic senators, with an IQR of 12 years compared to an IQR of 10 years for Republican senators.

3.57 a.   The median is 48. 50% of the southern states have more than 48 capital prisoners.

   b.   Q1 = 32, Q3 = 152, IQR = 120

   c.   The mean is 90.8.

   d.   The mean is pulled up by the really large numbers, such as Texas (243) and Florida (374).

   e.   The median is unaffected by outliers.

Column	n	Mean	Variance	Std. dev.	Std. err.	Median	Range	Min	Max	Q1	Q3
Capital Prisoners	15	90.8	11147.6	105.5822	27.261206	48	374	0	374	32	152

3.59

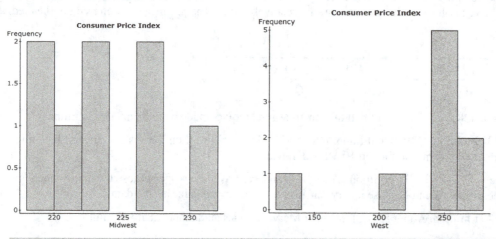

Column ‡	n ‡	Mean ‡	Variance ‡	Std. dev. ‡	Std. err. ‡	Median ‡	Range ‡	Min ‡	Max ‡	Q1 ‡	Q3 ‡
Midwest	8	223.6375	18.502679	4.301474	1.5208007	222.8	11.9	218.7	230.6	219.9	227.2
West	9	234.58889	1819.9511	42.660885	14.220295	244.6	141.4	128	269.4	240	258.6

Since the data for the West is left-skewed, use the median and IQR to compare the groups. The CPI for the West is higher than that of the Midwest (West median 244.6; Midwest median 222.8). There is more variability in the West CPI (West IQR 18.6; Midwest IQR 7.3). The West has one potentially low outlier.

3.61 a.   The balancing point of the histogram, the mean is approximately 80 millimeters.

   b. $\bar{x}$ = 1(60) + 8(70) + 8(80) + 5(90) + 3(100) + 1(100) + 1(120)/27 = 83.0 millimeters

   c.   It is an approximation because we used the left-hand side of the bin to estimate the data values contained in the bin.

Section 3.5: Using Boxplots for Displaying Summaries

3.63 a.   South, West, Northeast, Midwest.

     b.   Northeast, South, Midwest, West.

     c.   The South and the Northeast each have one state with a high-priced low poverty rate.

     d.   The Northeast has the least amount of variability (smallest IQR).

     e.   The IQR is better because is it not influenced by unusually high or low values and because the data is not symmetric.

3.65 a.   The NFL has the highest ticket prices and the most variability in ticket prices (highest median and greatest IQR). The MLB has the lowest ticket prices.

     b.   Hockey tickets tend to be more expensive than basketball tickets (higher median). Both sports have some unusually high-priced tickets, and hockey has more variability in ticket prices (greater IQR).

3.67 Explanations of reasoning will vary.

     a.   Histogram 1 is left-skewed, histogram 2 is roughly bell-shaped (not very skewed), and histogram 3 is right-skewed.

     b.   Histogram 1 goes with Boxplot C, since the boxplot is left-skewed.
Histogram 2 goes with Boxplot B, since the boxplot is not very skewed.
Histogram 3 goes with Boxplot A, since the boxplot is right-skewed.

3.69 The maximum value (756) is greater than the upper fence (237 + 1.5(237 - 63) = 498) and is considered a potential outlier. The minimum value (1) is not less than the lower fence.

3.71 The whiskers are drawn to the upper and lower fences or to the maximum and minimum values if there are no potential outliers in the data set. There are no values less than the lower fence, so the left whisker is drawn to the minimum. Since the maximum value (128.016) is beyond the upper fence (33.223 + 1.5(33.223 - 8.526) = 70.3) the right whisker would be drawn to the upper fence (70.3).

3.73 The IQR is 90 - 78 = 12. 1.5 * 15 = 18, so any score below 78 - 18 = 60 is a potential outlier. We can see that there is at least one potential outlier (the minimum score of 40), but we don't know how many other potential outliers there are between 40 and 60. Therefore, we don't know which point to draw the left-side whisker to.

Chapter Review Exercises

3.75 a.   The median is 41.05 cents/gallon. 50% of the southern states have gas taxes greater than 41.05 cents/gallon.

     b.   The middle 50% of the southern states have gas taxes in a range of 11.35 cents/gallon.

     c.   The mean is 43.6 cents/gallon.

     d.   The data may be right-skewed.

Column	n	Mean	Variance	Std. dev.	Std. err.	Median	Range	Min	Max	Q1	Q3
Gas Tax (cents/gallons)	16	43.6	40.749333	6.3835204	1.5958801	41.05	19.8	35.2	55	38.85	50.2

3.77 The 5 p.m. class did better, typically; both the mean and the median are higher. Also, the spread (as reflected in both the standard deviation and the IQR) is larger for the 11 a.m. class, so the 5 p.m. class has less variation. The visual comparison is shown by the boxplots. Both distributions are slightly left-skewed. Therefore, you can compare the means and standard deviations or the medians and IQRs.

The visual comparison is shown by the boxplots. Both distributions are slightly left-skewed. Therefore, you can compare the means and the standard deviations *or* the medians and the IQRs.

```
Minitab Statistics
Variable  N    Mean   Median  StDev  Min   Max    Q1  Q3
11am      15   70.73  72.5    19.84  39    100    53  86
5pm       19   84.78  86.5    11.95  64.5  104.5  73  94
```

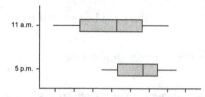

3.79 The graph is bimodal, with modes around 65 inches (5 feet 5 inches) and around 69 inches (5 feet 9 inches). There are two modes because men tend to be taller than women.

3.81 Students should provide histograms for males and females; because of lack of symmetry, they should compare medians and IQRs. Both groups typically watch about the same amount and have similar variability of lack of symmetry they should compare medians and IQRs.

3.83 a. The mean is approximately 1000 calories.

b. An estimate of the standard deviation is (2200 – 100)/6 = 350 calories.

3.85 Answers will vary.

3.87 Answers will vary.

3.89 Students should make a histogram for western states and southern states. Because the distributions are not symmetric, students should compare the medians and the IQRs. The western states tend to have a higher percentage of the population with a bachelor's degree. The southern states tend to have more variability.

3.91 a. Since the distributions are slightly right-skewed, compare the medians.

b. Football ticket prices tend to be higher than hockey tickets. Football tickets also show more variation in price. (IQR: Football $84, Hockey $67).

3.93 a. 80 + 2(4) = 88

b. 80 – 1.5(4) = 74

3.95 The *z*-score for the SAT of 750 is (750 – 500)/ 100 = 2.5, and the *z*-score for the ACT of 28 is (28-21)/5 = 1.4. The score of 750 is more unusual because its *z*-score is farther from 0.

3.97 a. The distribution is right-skewed.

b. Because the data are right skewed, the mean would be greater than the median.

c. The majority of ticket prices would be less than the mean price.

3.99 Answers will vary. "Who ran faster, grade 11 or grade 12 boys?" "Which group had the most consistent times, grade 11 or grade 12 boys?"

3.101 Answers vary. Possible answers include "Which buildings are typically taller: those made of concrete or those made of steel?" "Is there more variability in the heights of building constructed in the 2000s or before 2000?"

Chapter 4: Regression Analysis: Exploring Associations between Variables

Section 4.1: Visualizing Variability with a Scatterplot

4.1 The critical reading score is a somewhat better predictor because the vertical spread is less, suggesting a more accurate prediction of GPA.

4.3 The trend appears roughly linear and positive up to about age 24, and then it starts to curve.

4.5 There is very little trend. It appears that number of credits acquired is not associated with GPA.

4.7 The trend is positive. Students with more sisters tended to have more brothers. This trend makes sense, because large families are likely to have a large number of sons and a large number of daughters.

4.9 There is a slightly negative trend. The negative trend suggests that the more hours of work a student has, the fewer hours of TV the student tends to watch. The person who works 70 hours appears to be an outlier because that point is separated from the other points by a large amount.

4.11 There is a slight negative trend that suggests that older adults tend to sleep a bit less than younger adults. Some may say there is no trend.

Section 4.2: Measuring Strength of Association with Correlation

4.13 a. You should not find the correlation because the trend is not linear.

b. You may find the correlation because the trend is linear.

4.15 The correlation coefficient is positive since the graph shows an upward trend.

4.17 Since it has a stronger positive association, 0.767 goes with graph A.
Since it has a weaker positive association, 0.299 goes with graph B.
Since it is the only graph with a negative association, –0.980 goes with graph C.

4.19 Since it has a stronger positive association, 0.87 goes with graph A.
Since it is the only graph with a negative association, –0.47 goes with graph B.
Since it has a weaker positive association, 0.67 goes with graph C.

4.21 R (correlation coefficient) = 0.68518783

a. $r = 0.69$

b. $r = 0.69$; the correlation coefficient stays the same.

c. $r = 0.69$. Adding a constant to all y-values does not change the value of r.

d. $r = 0.69$; the correlation coefficient stays the same.

4.23 The correlation would not change. The correlation does not depend on which variable is the predictor and which is the response.

4.25 The correlation is 0.904. The professors that have high overall quality scores tend to also have high easiness scores.

4.27 Higher gym usage is associated with higher GPA.

Section 4.3: Modeling Linear Trends

4.29 a. The independent variable is median starting salary, and the dependent variable is median mid-career salary.

b. Salary distributions are usually skewed. Medians are therefore a more meaningful measure of center.

   c.   Between $110,000 and $120,000.

   d.   Mid-Career $= -7699 + 1.989$ Starting $= -7699 + 1.989(60,000) = \$111,641$

   e.   Answers will vary. The number of hours worked per week, the amount of additional education required, gender, and the type of career are all factors that might influence mid-career salary.

4.31 a.   The median pay for women is about $690 when pay for men is $850.

     b.   predicted women $= -62.69 + 0.887(850) = 691.26$.

     c.   The slope is 0.887. Each additional dollar in men's' pay is associated with an average increase of $0.887 in the median womens' pay.

     d.   The *y*-intercept is -62.69. It is not appropriate to interpret it in this context because the median income for men (*x*) cannot be zero.

4.33 a.   Predicted Armspan $= 16.8 + 2.25$ Height

     b. $b = r\dfrac{s_y}{s_x} = 0.948\left(\dfrac{8.10}{3.41}\right) = 2.25$

     c. $a = \bar{y} - b\bar{x} = 159.86 - 2.25(63.59) = 16.8$

     d.   Predicted Armspan $= 16.8 + 2.25$ Height $= 16.8 + 2.25(64)$ 160.8, or about 161 cm

4.35 a.   Predicted Armspan $= 6.24 + 2.515$ Height (Rounding may vary.)

     b.   Minitab: slope = 2.51, intercept = 6.2

         StatCrunch: slope = 2.514674, intercept = 6.2408333

         Excel: slope = 2.514674, intercept = 6.240833

         TI-84: slope = 2.514673913, intercept = 6.240833333

4.37 The association for the women is stronger because the correlation (*r*-value) is closer to 1.

4.39 a.   Based on the scatterplot there is not a strong association between these two variables.

     b.   The numerical value of the correlation would be close to zero because there is not an association between these variables.

     c.   Since there is not an association between these variables, we cannot use singles percentage to predict doubles percentage.

4.41 Explanations will vary.

	x	*y*
a.	odometer reading	price
b.	household size	water bill
c.	time spent in gym	weight

4.43 a.   The higher the percentage of smoke-free homes in a state, the lower the percentage of high school students who smoke tends to be.

     b.   Predicted Pct. Smokers $= 56.32 - 0.464$ (Pct. Smoke Free) $= 56.32 - 0.464(70) = 23.84$, or about 24%

4.45 a.   As driver age increases, insurance prices decrease but then begin to increase again at around 65 years of age. Younger drivers and older drivers tend to have more accidents so they are charged more for insurance.

     b.   It would not be appropriate to do a linear regression analysis on these data because the data do not follow a linear trend.

4.47   The answers follow the guidance on page 209.

   1:

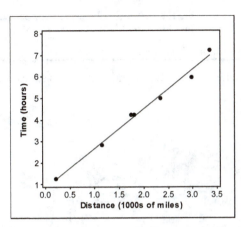

   2:   The linear model is appropriate. The points suggest a straight line.

   3:   Predicted Time = 0.8394 + 1.838 Distance

   Regression Analysis: Time (hours) versus Distance (1000s of miles)
   The regression equation is
   Time (hours) = 0.8394 + 1.838 Distance (1000s of miles)

   4:   Each additional thousand miles takes, on average, about 1.84 more hours (or 110 minutes) to arrive.

   5:   The additional time shown by the intercept might be due to the time it takes for the plane to taxi to the runway, delays, the slower initial speed, and similar delays in the landing as well. The time for this appears to be about 0.84 hours (or 50 minutes).

   6:   Predicted Time = 0.8394 + 1.838(3) = 0.8394 + 5.514 = 6.35 hours. The predicted time for a flight from Boston to Seattle is about 6.35 hours.

4.49   a.   The slope and correlation will be positive: The more population, the more millionaires there tend to be.

       b.

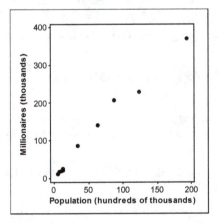

       c. $r = 0.992$

   Correlations: Millionaires (thousands), Population (hundreds of thousands)
   Pearson correlation of Millionaires (thousands) and Population (hundreds of
       Thousands) = 0.992
   ---

   d.   The slope is 1.9. For each additional hundred thousand in the population, there is an additional 1.9 thousand millionaires.

Regression Analysis: Millionaires (thousands) versus Population (hundreds of thousands)
The regression equation is
Millionaires (thousands) = 6.30 + 1.92 Population (hundreds of thousands)

---

   e.   Do not focus on the intercept, because it does not make sense to look for millionaires in states with no people.

4.51 a.

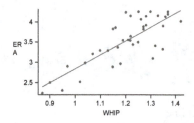

   b.   ERA = -0.578 + 3.436 WHIP

   c.   Each additional point in WHIP rating is associated with an average increase of 3.436 in ERA.

   d.   It would be inappropriate to interpret the y-intercept because there are no pitchers with a 0 WHIP rating.

Section 4.4: Evaluating the Linear Model

4.53 a.   Influential points are outliers in the data that can have large effects on the regression line. When influential points are present in the data, do the regression and correlation with and without these points and comment on the difference.

   b.   The coefficient of determination is the square of the correlation coefficient; it measures the percentage of variation in the y-variable that is explained by the regression line.

   c.   Extrapolation means using the regression equation to make predictions beyond the range of the data. Extrapolation should not be used.

4.55 Older children have larger shoes and have studied math longer. Large shoes do not cause higher grades. Both are affected by age.

4.57 $(0.67)^2 = 0.4489$, so the coefficient of determination is about 45%. Therefore, 45% of the variation in weight can be explained by the regression line.

4.59 Part of the poor historical performance could be due to chance, and if so, regression toward the mean predicts that stocks turning in a lower than average performance should tend to perform closer to the mean in the future. In other words, they might tend to increase.

4.61 a.   The slope of –2099 means that the salary is $2099 less for each year later that the person was hired, or $2099 more for each year earlier.

   b.   The intercept ($4,255,000) would be the salary for a person who started in the year 0, which is inappropriate (and ridiculous).

4.63 a.

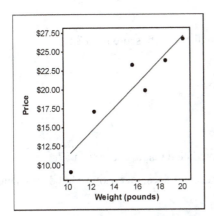

b. $r = 0.933$; A positive correlation suggests that larger turkeys tend to have a higher prices.

Correlations: Weight (pounds), Price
Pearson correlation of Weight (pounds) and Price = 0.933

---

c. Predicted Price $= -4.49 + 1.57$ Weight

Regression Analysis: Price versus Weight (pounds)
The regression equation is
Price = - 4.49 + 1.57 Weight (pounds)

---

d. The slope of 1.57 means that for each additional pound, the price goes up by $1.57. The interpretation of the intercept is inappropriate, because it is not possible to have a turkey that weighs 0 pounds.

e. $r = -0.375$; Predicted Price $= 26.87 - 0.533$ Weight. The negative correlation and slope imply that the bigger the turkey, the less it costs! The 30-pound free turkey was an influential point, which really changed the results.

Correlations: Weight (pounds), Price
Pearson correlation of Weight (pounds) and Price = -0.375
Regression Analysis: Price versus Weight (pounds)
The regression equation is
Price = 26.87 - 0.553 Weight (pounds)

---

f. $r^2 = 0.87$. 87% of the variation in turkey price is explained by weight.

4.65 a. Positive.

b. Slope $= 0.327$; each additional $1 in teacher pay is associated with an increase in per pupil spending of 0.327.

c. The y-intercept is -5922; it would not be appropriate to interpret the y-intercept because there is no state an average teacher salary of $0.

d. $-5922 + 0.327(60000) = \$13,698$.

4.67 a. Since the y-values decrease as the x-values increase, there is a negative correlation.

b. For each additional hour of work, the score tended to go down by 0.48 point.

c. A student who did not work would expect to get about 87 on average.

4.69 There is a stronger association between home runs and strikeouts ($r = 0.64$ compared to $r = -0.09$).

4.71 a.
Dependent Variable: 4th Grade Math
Independent Variable: 4th Grade Reading
4th Grade Math = 33.027886 + 0.6961738 4th Grade Reading
Sample size: 20
R (correlation coefficient) = 0.84877382

$r = 0.85$;
Predicted Math Score = 33.03 + 0.70 (Reading score)
Predicted Math Score = 33.03 + 0.70(70) = 82.03. The predicted math score is an 82.

b.

Dependent Variable: 4th Grade Reading
Independent Variable: 4th Grade Math
4th Grade Reading = -15.334137 + 1.0348235 4th Grade Math
Sample size: 20
R (correlation coefficient) = 0.84877382

$r = 0.85$;
Predicted Reading Score = −15.33 + 1.03(70) = 57. The predicted reading score is a 57.

c.   Changing the choice of the dependent and independent variables does not change r but does change the regression equation.

4.73   The answers follow the guided steps.

1:    a. $b = r \cdot \dfrac{s_{final}}{s_{midterm}} = 0.7 \cdot \dfrac{10}{10} = 0.7$

b. $a = \bar{y} - b\bar{x} = 75 - 0.7(75) = 75 - 52.5 = 22.5$

c.   Predicted Final = 22.5 + 0.7 Midterm

2:    Predicted Final = 22.5 + 0.7 Midterm = 22.5 + 0.7(95) = 22.5 + 66.5 = 89

3:    The score of 89 is lower than 95 because of regression toward the mean.

Chapter Review Exercises

4.75   a. $r = 0.941$; Predicted Weight = −245 + 5.80 Height

Regression Analysis: Weight (pounds) versus Height (inches)
The regression equation is
Weight (pounds) = - 245 + 5.80 Height (inches)
Correlations: Height (inches), Weight (pounds)
Pearson correlation of Height (inches) and Weight (pounds) = 0.941

b.

Height (cm)	Weight (kg)
$60(2.54) = 152.40$	$\dfrac{105}{2.205} = 47.6191$
$66(2.54) = 167.64$	$\dfrac{140}{2.205} = 63.4921$
$72(2.54) = 182.88$	$\dfrac{185}{2.205} = 83.9002$
$70(2.54) = 177.80$	$\dfrac{145}{2.205} = 65.7596$
$63(2.54) = 160.02$	$\dfrac{120}{2.205} = 54.4218$

c.   The correlation between height and weight is 0.941. It does not matter whether you use inches and pounds or centimeters and kilograms. A change of units does not affect the correlation because it has no units.

Correlations: Height (cm), Weight (kg)
Pearson correlation of Height (cm) and Weight (kg) = 0.941

d. The equations are different, using different units will result in different results.

Predicted Weight Pounds = –245 + 5.80 Height (inches)

Predicted Weight Kilograms = –11 + 1.03 Height (centimeters)

Regression Analysis: Weight (lb) versus Height (in)
The regression equation is
Weight (pounds) = - 245 + 5.80 Height (inches)
Regression Analysis: Weight (kg) versus Height (cm)
The regression equation is
Weight (kg) = - 111 + 1.03 Height (cm)

4.77 a.

Dependent Variable: Calories
Independent Variable: Carbs (grams)
Calories = 3.2622227 + 10.806543 Carbs (grams)
Sample size: 48
R (correlation coefficient) = 0.85821218

$r = 0.86$; Predicted Calories = 3.26 + 10.81(Carbs); Slope = 10.81; Each additional gram of carbohydrates is associated with an increase of 10.81 calories
Predicted Calories = 3.26 + 10.81(55) = 597.81 calories

b.

Dependent Variable: Calories
Independent Variable: Sugars (grams)
Calories = 198.73601 + 32.745312 Sugars (grams)
Sample size: 48
R (correlation coefficient) = 0.79253604

$r = 0.79$; Predicted Calories = 198.74 + 32.75(sugars)
Predicted Calories = 198.74 + 32.75(10) = 526.24 calories

c. While both are fairly good, the number of carbs is a better predictor ($r = 0.86$ compared to $r = 0.79$).

4.79 a. There were no women taller than 69 inches, so the line should stop at 69 inches to avoid extrapolating.

b. Men who are the same height as women wear shoes that are, on average, larger sizes.

c. The mean increase in shoe size based on height is the same for men and women.

4.81 Among those who exercise, the effect of age on weight is less. An additional year of age does not lead to as great an increase in the average weight for exercisers as it does for non-exercisers.

4.83 a.

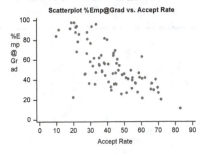

There seems to be a linear trend. The less-selective law schools also tend to have a lower employment rate at graduation.

b. I. % Employed = 97.59 - 1.03 Acceptance Rate
II. Each additional percentage point in acceptance rate is associated with an average decrease of 1.03 percentage points in the rate of employment at graduation.
III. It would be inappropriate to interpret the y-intercept because there is no school with an acceptance rate of 0%.
IV. $r^2 = 52\%$. 52% of the variation in employment rate at graduation can be explained by the acceptance rate.

V. The regression equation predicts an employment rate of about 32.6% for a school with an acceptance rate of 50%.

4.85   A comparison of scatterplots shows a stronger association between calories and fat compared with calories and carbohydrates. The correlation coefficient for fat and calories is $r = 0.82$ while the correlation coefficient for carbohydrates and calories is 0.39. Fat is a better predictor of the number of calories in these snack foods.

4.87

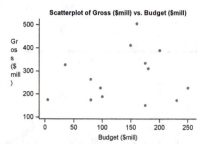

It is not appropriate to fit a linear regression model, because the trend is not linear. However, we can see that some big-budget films didn't do as well compared to some lower budget films. The one point at the top, Wonder Women, had the highest gross but did not have the highest budget.

4.89   a.   The positive trend shows that the more stories there are, the taller the building tends to be.

   b.   Predicted Height $= 115.4 + 12.85(100) = 115.4 + 1285 = 1400.4$, or about 1400 feet

   c.   The slope of 12.85 means that buildings with one additional story tend to have an average of 12.35 feet of additional height.

   d.   Because there are no building with 0 stories, the interpretation of the intercept is not appropriate.

   e.   About 71% of the variation in height can be explained by the regression, and about 29% is not explained.

4.91   Answers will vary.

x1	y1
2	5
3	6
4	7
5	8

Correlations: x1, y1
```
Pearson correlation of x1 and y1 = 1.000
```

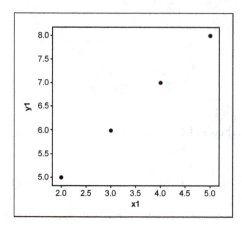

4.93   Answers will vary.

Dataset 1		Dataset 2	
x1	y1	x2	y2
5	8	5	8
6	7	6	7
7	6	7	6
8	5	8	5
		1	1

Correlations: x1, y1
Pearson correlation of x1 and y1 = -1.000
Correlations: x2, y2
Pearson correlation of x2 and y2 = 0.658

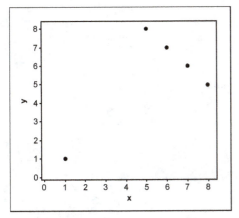

4.95   The trend is positive. In general, if one twin has a higher-than-average level of education, so does the other twin. The point that shows one twin with 1 year of education and the other twin with 12 years is an outlier. (Another point showing one twin with 15 years and the other with 8 years is a bit unusual, as well.)

4.97   There appears to be a positive trend. It appears that the number of hours of homework tends to increase slightly with enrollment in more units.

4.99   Linear regression is not appropriate because the trend is not linear, it is curved.

4.101 The cholesterol going down might be partly caused by regression toward the mean.

Chapter 5: Modeling Variation with Probability

Section 5.1: What Is Randomness?

5.1 Answers may vary with different random numbers.

   a.   5   5   1   8   5   7   4   8   3   4

   b.   T   T   H   T   T   T   H   T   H   H

   c.   4/10 = 40% were heads.

5.3 This is a theoretical probability, because it is not based on an experiment.

5.5 This is an empirical probability, because it is based on an experiment.

Section 5.2: Finding Theoretical Probabilities

5.7   a.   The seven equally likely outcomes are Town, Wu, Hein, Lee, Marland, Penner, and Holmes.

      b.   The probability that the new patient will be assigned to a female doctor is $4/7$, or about 57.1%.

      c.   The probability that new patient will be assigned to a male doctor is $3/7$, or about 42.9%.

      d.   Yes, the event "male doctor" is equivalent to the event "not a female doctor".

5.9   a.   0.26 can be the probability of an event.

      b.   -0.26 cannot be a probability because it is negative.

      c.   2.6 cannot be the probability of an event since 2.6 is greater than 1.

      d.   2.6% can be the probability of an event.

      e.   26 cannot be the probability of an event since 26 is greater than 1.

5.11   a. $P(\text{a heart}) = 13/52$, or 1/4 or 25%

      b. $P(\text{a red card}) = 26/52$, or 1/2 or 50%

      c. $P(\text{an ace}) = 4/52$, or 1/13 or about 7.7%

      d. $P(\text{a face card}) = 12/52$, or 3/13 or about 23%

      e $P(\text{a three}) = 4/52$, or 1/13 or about 7.7%

5.13   a. $P(\text{guessing correctly}) = 1/2$, or 50%

      b. $P(\text{guessing incorrectly}) = 1/2$, or 50%

5.15   a. $1+4+6+4+1 = 16$ outcomes

      b.   i. $1/16 = 6.25\%$

          ii. $4/16 = 25\%$

          iii. $5/16 = 31.25\%$

5.17 $P(\text{Friday OR Saturday OR Sunday}) = P(\text{Friday}) + P(\text{Saturday}) + P(\text{Sunday})$

$$= \frac{1}{7} + \frac{1}{7} + \frac{1}{7}$$

$$= \frac{3}{7}, \text{ or about } 42.86\%$$

5.19 a. i. P(college graduate) $= 250/500 = 50\%$

ii. P(vacation) $= 335/500 = 67\%$

iii. P(college graduate and vacation) $= 200/500 = 40\%$

b. P(CG Vacation) $= 200/250 = 80\%$, P(NCG Vacation) $= 135/250 = 54\%$

College graduates were more likely to take vacation.

5.21 a. P(Republican) $= 278/1028 = 27\%$

b. P(Democrat) $= 308/1028 = 30\%$

c. P(Republican or Democrat) $= (278+308)/1028 = 57\%$

d. "Republican" and "Democrat" are mutually exclusive because a person cannot be both a Democrat and a Republican at the same time.

e. " Republican" and "Democrat" are not complementary because "not Republican" includes Democrats and Independents and "Democrat" are not mutually exclusive because a person can be in both groups at the same time.

5.23 a. P(Man or About the same) $= (280+259-174) = 84.9\%$

b. P (Mand and About the same) $= 174/430 = 40.5\%$

5.25 a. i. Not mutually exclusive; some people have travelled to both Mexico and Canada

ii. Mutually exclusive; a person cannot be single and married at the same time.

b. Answers will vary.

5.27 Answers vary. Being a college graduate or not being a college graduate are mutually exclusive. Saying Yes and No are mutually exclusive. Naming any two rows (or two columns) gives you mutually exclusive events.

5.29 a. The numbers on a die that are even or greater than 4 are 2, 4, 5, 6 so the probability is $4/6 = 66.7\%$.

b. The numbers on a die that are odd and or less than 3 are 1, 2, 3, and 5, so the probability is $4/6 = 66.7\%$.

5.31 a. Mutually exclusive

$$P(A \text{ OR } B) = P(A) + P(B)$$
$$= 0.18 + 0.25$$
$$= 0.43, \text{ or } 43\%$$

b. Mutually exclusive

$$P(A \text{ OR } B \text{ OR } C) = P(A) + P(B) + P(C)$$
$$= 0.18 + 0.25 + 0.37$$
$$= 0.80, \text{ or } 80\%$$

c. Complement

$$P\left(\text{lower than C}\right) = 1 - P\left(\text{A OR B OR C}\right)$$
$$= 1 - 0.80$$
$$= 0.20, \text{ or } 20\%$$

5.33 Since the total of all responses must equal 100%, 100% - (42% + 23%) = 35%.

5.35 a. 17% + 15% = 32%

b. 20% + 29% = 49%; 49% * 1200 = 588 people

5.37 a. $P\left(\text{two Ups}\right) = P\left(UU\right)$
$$= P\left(U\right) \times P\left(U\right)$$
$$= \left(0.6\right)\left(0.6\right)$$
$$= 0.36, \text{ or } 36\%.$$

b. $P\left(\text{one Up}\right) = P\left(\text{UD or DU}\right)$
$$= P\left(UD\right) + P\left(DU\right)$$
$$= \left(0.6\right)\left(0.4\right) + \left(0.4\right)\left(0.6\right)$$
$$= 0.48, \text{ or } 48\%.$$

c. $P\left(\text{at least one Up}\right) = 1 - P\left(\text{no Ups}\right)$
$$= 1 - P\left(DD\right)$$
$$= 1 - \left(0.4\right)\left(0.4\right)$$
$$= 0.84, \text{ or } 84\%.$$

d. $P\left(\text{at most one Up}\right) = 1 - P\left(\text{two Ups}\right)$
$$= 1 - P\left(UU\right)$$
$$= 1 - \left(0.6\right)\left(0.6\right)$$
$$= 0.64, \text{ or } 64\%.$$

5.39 a. $P\left(\text{More than 8 mistakes}\right) = 1 - \left(P\left(\text{fewer than 3}\right) + P\left(\text{3 to 8 mistakes}\right)\right)$
$$= 1 - \left(0.48 + 0.30\right)$$
$$= 0.22, \text{ or } 22\%$$

b. $P\left(\text{3 or more mistakes}\right) = P\left(\text{3 to 8 mistakes}\right) + P\left(\text{more than 8 mistakes}\right)$
$$= 0.30 + 0.22$$
$$= 0.52, \text{ or } 52\%$$

c. $P\left(\text{at most 8 mistakes}\right) = P\left(\text{fewer than 3}\right) + P\left(\text{3 to 8 mistakes}\right)$
$$= 0.48 + 0.30$$
$$= 0.78, \text{ or } 78\%$$

d. The events in parts a and c are complementary because "at most 8 mistakes" means from 0 mistakes up to 8 mistakes. "More than 8 mistakes" means 9, 10, up to 12 mistakes. Together, these mutually exclusive events include the entire sample space.

Section 5.3: Associations in Categorical Variables

5.41 a.   i. P (has been about right I female)

  b. $34/100 = 34\%$

5.43 a. $319/550 = 58\%$

  b. $195/500 = 39\%$

  c. $(514 + 550 - 319)/1050 = 745/1050 = 71\%$ $(514 + 550 - 319)/1050 = 745/1050 = 71\%$

5.45 They are associated. Tall people are much more likely to play professional basketball than are short people. To say it another way, basketball players are much more likely to be tall than are those who don't play basketball.

5.47 Eye color and gender are independent because eye color does not depend on gender.

5.49 They are not independent because the probability of saying "Hasn't Gone Far Enough" if the person is female is $57/100 = 57\%$ whereas the overall probability of saying "Hasn't Gone Far Enough" is $99/200 = 49.5\%$ and these are not the same.

5.51 The answers follow the guided steps.

  1:

	Man	Woman	Total
Right-handed	18	42	60
Left-handed	12	28	40
Total	30	70	100

  2:    $P(\text{Right}) = \dfrac{60}{100} = \dfrac{3}{5}$, or 60%

  3:    $P(\text{Right} \mid \text{Male}) = \dfrac{P(\text{Right AND Male})}{P(\text{Male})}$

  $= \dfrac{\dfrac{18}{100}}{\dfrac{30}{100}}$

  $= \dfrac{18}{30} = \dfrac{3}{5}$, or about 60%

  4:    The variables are independent because the probability of having the right thumb on top given that a person is a man is equal to the probability that a person has the right thumb on top (for the whole data set).

5.53 a.

	Local TV	Network TV	Cable TV	Total
Men	66	48	58	172
Women	82	54	56	192
Total	148	102	114	364

  b.   P(cable $\mid$ male) $= 58/172 = 33.7\%$; P(cable) $= 114/364 = 331.3\%$. Since these probabilities are not equal, the events are not independent.

5.55 a. $1/8$

  b. $1/8$

5.57   They are the same. Both probabilities are $\left(\dfrac{1}{6}\right)^5 = \dfrac{1}{1776}$.

5.59   a.   P(vacation and vacation) = (0.62)(0.62) = 38.4%

       b.   P no vacation and no vacation) = (0.38)(0.38) = 14.4% (since vacation = .62, no vacation would be
            1 - .62 = .38)

       c.   P(at least one took vacation) = 100% – (no one took vacation) so 100% - 14.4% = 85.6%
            (The complement to at least one of something would be none.)

5.61 $P(\text{have C AND test pos}) = P(\text{have C}) \times P(\text{test pos} \mid \text{have C})$
$$= (0.00008)(0.84)$$
$$= 0.000067, \text{ or about } 0.0067\%$$

Section 5.4: Finding Empirical and Simulated Probabilities

5.63   a.   Trials, 2, 3 and 4 had at least one 6.

       b.   The empirical probability is $3/5$ or 0.6.

5.65   a. and b. are both valid methods because the probability of a correct choice is $1/3$.

       c.   is invalid because the probability of the correct choice is $3/10$.

5.67   a.   The random numbers represent the following outcomes: HHHTH HTHHT TTHTH HTTTH.

       b.   The simulated probability of getting heads is $11/20 = 55\%$, which is close to the theoretical probability
            50%.

       c.   The theoretical probability is 50%, so in this case the simulated probability is likely to be close to the
            theoretical probability.

5.69   Histogram B was for 10,000 rolls because it has nearly a flat top. In theory, there should be the same number
       of each outcome, and the Law of Large Numbers says that the one with the largest sample should be closest to
       the theory.

5.71   The proportion should get closer to 0.5 as the number of flips increases.

5.73   Betty and Jane are betting more times (100 times), so they are more likely to end up with each having about
       half of the wins, compared to Tom and Bill. The Law of Large Numbers says that the more times you try an
       experiment, the closer the experimental proportion comes to the theoretical proportion (50%). The graph
       shows that the proportion of wins settles down to about 50% by 100 trials. But at 10 trials, the percentage of
       wins has not settled down and will vary quite a bit. Using the graph for this one simulation, it looks like one of
       the men has won substantially more than 50% of the trials (after 10 trials).

5.75   You are equally likely to get heads or tails (assuming the coin is fair) because the coin's results are
       independent of each other—that is, the coin does not "keep track" of its past.

5.77   The probability of selecting a digit from 0 to 5 is $6/10$, or 60%, so it does not represent the probability we
       wish to simulate.

5.79   a.   You could use the numbers 1, 2, 3, and 4 to represent the outcomes and ignore 0 and 5–9, but answers to
            this will vary.

       b.   The empirical probabilities will vary. The theoretical probability of getting a 1 is 1/4; remember that the
            die is four-sided.

Chapter Review Exercises

5.81 $P(\text{Dissatisfied}) = \dfrac{627}{1012} = 62.0\%$

5.83   a.   Gender and shoe size are associated, because men tend to wear larger shoe sizes than women.

   b.   Win/loss record is independent of the number of cheerleaders. The coin does not know how many cheerleaders there are or have an effect on the number of cheerleaders.

5.85   a. $P(\text{both support}) = P(\text{man supports}) \times P(\text{woman supports})$
$$= (0.55)(0.43)$$
$$= 0.2365, \text{ or about } 24\%.$$

   b. $P(\text{neither supports}) = P(\text{man does not support}) \times P(\text{woman does not support})$
$$= (0.45)(0.57)$$
$$= 0.2565, \text{ or about } 26\%.$$

   c. $P(\text{only one supports}) = 1 - [P(\text{both support}) + P(\text{neither supports})]$
$$= 1 - (0.2365 + 0.2565)$$
$$= 0.5070, \text{ or about } 51\%.$$

   d. $P(\text{at least one supports}) = 1 - P(\text{neither supports})$
$$= 1 - 0.2565$$
$$= 0.7435, \text{ or about } 74\%.$$

5.87   a. $P(\text{both have used online dating}) = P(\text{first used online dating}) \times P(\text{second used online dating})$
$$= (0.27)(0.27)$$
$$= 0.0729, \text{ or about } 7.3\%.$$

   b.   Because the adults are Facebook friends, they may have similar attitudes regarding online dating.

5.89   a. $P(\text{both born on Monday}) = P(\text{Alicia born on Monday}) \times P(\text{David born on Monday})$
$$= \left(\frac{1}{7}\right)\left(\frac{1}{7}\right)$$
$$= \frac{1}{49}, \text{ or about } 2.0\%.$$

   b. $P(\text{Alicia OR David born on Monday}) = P(\text{Alicia}) + P(\text{David}) - P(\text{both on Monday})$
$$= \frac{1}{7} + \frac{1}{7} - \frac{1}{49}$$
$$= \frac{13}{49}, \text{ or about } 26.5\%.$$

5.91   a. $(0.61)(2500) = 1525$

   b. $(0.31)(2500) = 775$

   c. $(0.36)(2500) = 900$

   d.   The responses are mutually exclusive because those surveyed can fall into only one of these categories.

5.93   Using the table:

a. $P(\text{Watch at least some Baseball}) = 1 - (.56) = .44$ or 44%

b. $(0.56)(400) = 224$, we would expect 224 would not watch any baseball this season.

5.95   Democrats: $(0.72)(1500) = 1080$, Republicans: $(0.36)(1500) = 540$ 1080 democrats and 540 Republicans felt that colleges and Universities have a positive effect on the country.

5.97   a. $P(0 \text{ heads}) = 1/4$, or 25%

b. $P(1 \text{ head}) = 2/4 = 1/2$, or 50%

c. $P(2 \text{ heads}) = 1/4$, or 25%

d. $P(\text{at least one head}) = 1 - P(\text{no heads}) = 1 - 1/4 = 3/4$, or 75%

e. $P(\text{no more than 2 heads}) = P(\text{two or fewer heads}) = 4/4 = 1$, or 100%

5.99   a.

	Democrats	Republicans	Independents	Total
Yes	300	153	134	587
No	100	147	66	313
Total	400	300	200	900

Using the table:

b.   P(democrat who said yes) = $300 / 900 = 33.3\%$

c.   P(republican who said no) = $147 / 900 = 16.3\%$

d.   P(no | republican) = $147 / 300 = 49\%$

e.   P(republican | no) = ($147 / 313 = 47\%$

f.   P(democrat or republican) = $700 / 900 = 77.8\%$

5.101   a.   P(both pass) = $(0.70)(0.70) = 49\%$

b.   P(only one passes) = $(0.70)(0.30) + (0.30)(0.70) = 42\%$

c.   P(neither pass) $(0.30)(0.30) = 9\%$

5.103   Recidivism and gender are not independent. If they were independent, the recidivism rates for men and women would be the same.

5.105   Answers vary.

5.107   The smaller hospital will have more than 60% girls born more often because, according to the Law of Large Numbers, there's more variability in proportions for small sample sizes. For the larger sample size ($n = 45$), the proportion will be more "settled" and will vary less from day to day. Over half the subjects in Tversky and Kahneman's study said that "both hospitals will be the same." But *you* didn't, did you?

5.109   a.

	Conservative Republican	Moderate/Liberal Republican	Moderate/Conserv. Democrat	Liberal Democrat	Total
Yes	65	77	257	332	731
No	368	149	151	88	756
Total	433	226	408	420	1487

b.   P(Conservative Republic) = $433 / 1487 = 29.1\%$

    c.   731/1487 = 29.1%

5.111 P(Conservative Republican and agrees) = 65 / 1487 = 4.4%

5.113 P(Liberal or Moderate/Conservative Democrat) = 408 + 420 / 1487 = 55.7%

    These events are mutually exclusive because a person cannot be in both categories.

5.115 P(Liberal Democrat or no) = (420 + 756 − 88) / 1487 = 1088 / 1487 = 73.2%

    These events are not mutually exclusive because a person can belong to both categories.

5.117 a.   332/420 = 79.0%

    b.   65/433 = 15.0%

    c.   65/731 = 8.9 %

5.119 a.   Not mutually exclusive because a person can belong to both categories.

    b.   Mutually exclusive because a person cannot belong to both categories.

5.121 a.   Trials 1 and 3 had at least 3 of the dice land on the same number.

    b.   The empirical probability is 2/5, or 0.4.

5.123 a.   The probability that the student will correctly guess an answer is 0.20 (1 out of 5). The probability of randomly selecting a 0 or a 1 is 2/10 = 1/5 = 0.20.

    b.

1	1	3	7	3		9	6	8	7	1
R	R	W	W	W		W	W	W	W	R

    c.   Yes. There were 3 correct.

    d.   W   W R W W W W R W W   No. The student scored only 2 correct.

    e.

R	W	W	R	W		W	W	W	W	R. 3 correct. Yes.
W	R	W	W	W		W	W	W	W	W 1 correct. No.

    f.   There were four trials, and two had a successful outcome. Thus, the empirical probability is $\frac{2}{4} = \frac{1}{2}$, or 50%.

5.125 a.   The action is to arrive at a light, which is either green or not. The probability of a success is 60%.

    b.   Answers will vary. Our method: Let the digits from 0 to 5 represent a green light, and let the digits from 6 to 9 represent yellow or red. (Any assignment that gives six digits to green and four digits to non-green will work.)

    c.   The event of interest is "get three out of three green."

    d.   A single trial consists of reading off three digits.

    e.   Outcomes (non-green are labeled R). Three greens in a row are underlined.

2 7 5	8 3 0	1 8 6	6 5 8	2 5 0	3 8 1	0 3 3	5 8 2	5 9 4	5 1 3
G R G	R G G	G R R	R G R	<u>G G G</u>	G R G	<u>G G G</u>	G R G	G R G	<u>G G G</u>
6 0 8	0 1 0	4 4 3	9 5 8	6 2 1	0 9 8	4 0 3	5 1 1	9 6 0	3 7 2
R G R	<u>G G G</u>	<u>G G G</u>	R G R	R G G	G R R	<u>G G G</u>	<u>G G G</u>	R R G	G R G

    Number of greens (with successful events in underlined)

    2 2 1 1 <u>3</u> 2 <u>3</u> 2 2 <u>3</u>

    1 <u>3</u> <u>3</u> 1 2 1 <u>3</u> <u>3</u> 1 2

    f.   Estimated probability: P(all three green) = 7/20, or 35%.

Chapter 6: Modeling Random Events: The Normal and Binomial Models

Section 6.1: Probability Distributions Are Models of Random Experiments

6.1 a. Continuous

b. Discrete

6.3 a. Continuous

b. Continuous

6.5 The table could also be presented vertically.

Number of Spots	1	2	4	5	6
Probability	0.10	0.20	0.20	0.20	0.10

6.7 $0.36 + 0.24 + 0.24 + 0.16 = 1$

Outcome	UU	UD	DU	DD
Probability	$0.6(0.6) = 0.36$	$0.6(0.4) = 0.24$	$0.4(0.6) = 0.24$	$0.4(0.4) = 0.16$

6.9 a. $0.36 + 0.24 + 0.24 + 0.16 = 1$

Number of Ups	0	1	2
Probability	0.16	$0.24 + 0.24 = 0.48$	0.36

b.

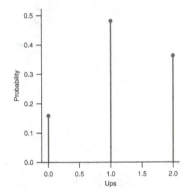

6.11 $(6-3)(0.2) = 3(0.2) = 0.60$, or 60%

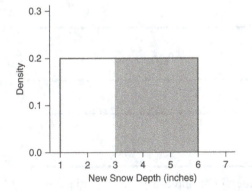

Section 6.2: The Normal Model

6.13 a.   ii. 95%

  b.   i. almost all

  c.   iii. 68%

  d.   iv. 50%

  e.   ii. 13.5%

6.15 The SAT scores that correspond to the *z*-scores are $500 - 3(100) = 200$, $500 - 2(100) = 300$, $500 - 1(100) = 400$, $500 + 1(100) = 600$, $500 + 2(100) = 700$, and $500 + 3(100) = 800$.

  a.   iii. 50%   (percentage above the mean)

  b.   iii. 68%   (percentage between *z*-scores of –1 and 1)

  c.   v. about 0%   (percentage above *z*-score of 3)

  d.   v. about 0%   (percentage below *z*-score of –3)

  e.   ii. 95%   (percentage between *z*-scores of –2 and 2)

  f.   v. 2.5%   (percentage between *z*-scores of 2 and 3, $(100\% - 95\%)/2 = 2.5\%$)

6.17 a. and b.
Use the output from (b), which shades the region above 67 inches. The percentage of college women with heights of above 67 inches is 21.1%.

6.19 a.   0.8708, or about 87%

  b. $1 - 0.8708 = 0.1292$, or about 13%

6.21 a.   0.9788 or about 98%

  b.   0.9599 or about 96%

  c.   0.8136 or about 81%

6.23 Graphs not provided.

  a. $1.000 - 1.000 = 0.000$

```
Cumulative Distribution Function
Normal with mean = 0 and standard deviation = 1
x   P( X <= x )
4      0.999968
```

  b. $1.000 - 1.000 = 0.000$

```
Cumulative Distribution Function
Normal with mean = 0 and standard deviation = 1
 x   P( X <= x )
10            1
```

  c. $1.000 - 1.000 = 0.000$

```
Cumulative Distribution Function
Normal with mean = 0 and standard deviation = 1
 x   P( X <= x )
50            1
```

  d.   The proportion to the right of *z* = 4.00 would be the largest of the three, and the proportion to the right of *z* = 50.00 would be the smallest.

  e.   Due to the symmetry of the Normal distribution, the area to the left of *z* = –10 would be equal to the area to the right of *z* = 10.

6.25   a. $z = \dfrac{170 - 150}{10} = \dfrac{20}{10} = 2$

b.   The Empirical Rule states that the area between -2 and 2 is 95%, so the area outside -2 and 2 must be 100% - 95% = 5%. This means half of the 5% must be below -2 and half must be above 2. Therefore, 2.5% must be above a standard score of 2. This corresponds with the probability an adult St. Bernard weighs more than 170 pounds.

c.   See graph below.

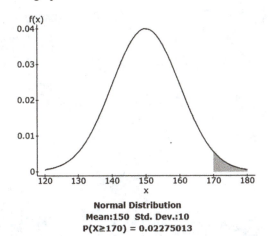

Normal Distribution
Mean:150 Std. Dev.:10
P(X≥170) = 0.02275013

d.   1: Almost all the values will lie between 3 standard deviations below the mean and 3 standard deviations above the mean, so almost all adult St. Bernard's will have weights between 150 ±3(10) or between 120 and 180 pounds.

2: 500 is the mean, and it belongs right below 0 because 0 is the mean of the standard Normal or the mean z-score.

6.27   About 1.2%.

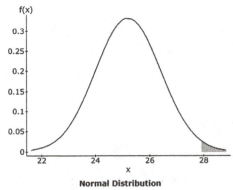

Normal Distribution
Mean:25.2 Std. Dev.:1.2
P(X≥27.9) = 0.01222447

6.29   About 69% of boys in this age group **will** have feet that are 24.6 to 28.8 cm long. The percentage that **will not** have feet in this range is 100% - 69% = 31%.

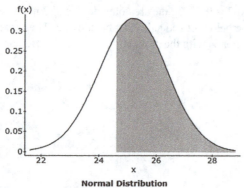

6.31   a.   About 13.4%.

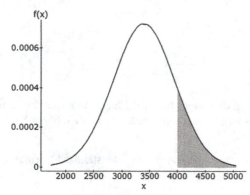

       b.   About 62.7%.

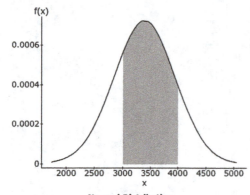

c. About 5.3%.

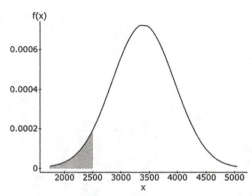

Normal Distribution
Mean:3390 Std. Dev.:550
P(X≤2500) = 0.05281171

6.33 a. About 72.8 %.

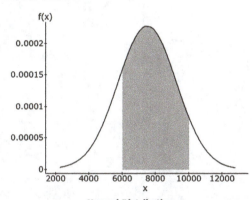

Normal Distribution
Mean:7500 Std. Dev.:1750
P(6000≤X≤10000) = 0.72775331

b. About 4.3%.

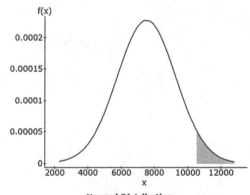

Normal Distribution
Mean:7500 Std. Dev.:1750
P(X≥10500) = 0.04323813

c.   About 4.3 %.

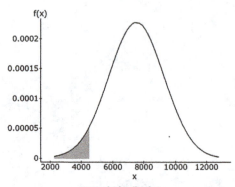

6.35 a.   About 33%.

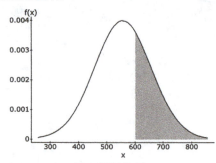

b.   About 15.6%.

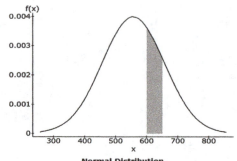

c.   A score of 720.

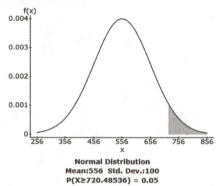

6.37 a.   About 86.7%.

b. $z = (89 - 71.4)/3.3 = 5.33$. About 0% of male have an arm span at least as long as 89 inches.

6.39 About 80% of the days in February have minimums of 32 °F or lower.

6.41 a.   Probability; about 4.8%.

b.   Measurement (inverse); 75 inches.

6.43 $z = 0.43$

6.45 a. $z = 0.56$

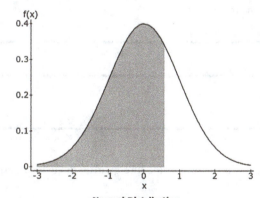

b. $z = -1.00$

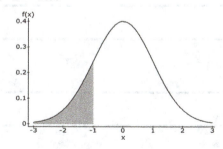

6.47 a.   About 7.7%.

b.   281 days.

c.   Almost 0%.

6.49 a.   94th percentile.

b.   A score of 517.

6.51 62.9 inches (5 feet 2.9 inches).

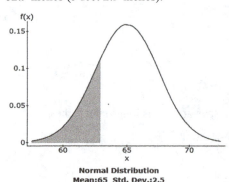

Normal Distribution
Mean:65 Std. Dev.:2.5
P(X≤62.895947) = 0.2

6.53 a.   The 75th percentile SAT score is 567.

Inverse Cumulative Distribution Function
```
Normal with mean = 500 and standard deviation = 100
P( X <= x )      x
     0.75   567.449
```

b.   The 25th percentile SAT score is 433.

Inverse Cumulative Distribution Function
```
Normal with mean = 500 and standard deviation = 100
P( X <= x )      x
     0.25   432.551
```

c. $567 - 433 = 134$

d.   The IQR is $567 - 433 = 134$ and the standard deviation is 100, so the interquartile range is larger.

6.55 a.   The girl is in the 77th percentile.

Cumulative Distribution Function
```
Normal with mean = 45 and standard deviation = 2
   x  P( X <= x )
46.5     0.773373
```

b.   She would be 66 inches (5 feet 6 inches) tall.

Inverse Cumulative Distribution Function
```
Normal with mean = 64 and standard deviation = 2.5
P( X <= x )      x
     0.77  65.8471
```

6.57 a.   The 90th percentile is 3.5 ounces.

Inverse Cumulative Distribution Function
```
Normal with mean = 3 and standard deviation = 0.4
P( X <= x )      x
     0.9  3.51262
```

b.   The 10th percentile is 2.5 ounces

```
Inverse Cumulative Distribution Function
Normal with mean = 3 and standard deviation = 0.4
P( X <= x )        x
         0.1   2.48738
```

Section 6.3: The Binomial Model (Optional)

6.59   Conditions
Two complementary outcomes: boy or girl
Fixed number of trials: 4 children
Same probability of success on each trial: 1/2 probability of a boy
All trials independent: Because there are no twins, the gender of each child is independent of the gender of the others.

6.61   Each trial has six possible outcomes (1, 2, 3, 4, 5, 6). In a binomial experiment, each trial only has two possible outcomes.

6.63   There is not a fixed number of trials because he is going to take as many free throws as he can in one minute

Note: The StatCrunch Binomial Calculator was used for the following calculations (6.65 – 6.76).

6.65   a. $b(20, 0.53, 12) = 0.1474$, or about 14.7%

b. $b(20, 0.47, 10) = 0.1699$, or about 17%

6.67   a. $b(10, 0.36, 4) = 0.242$ or about 24%

b. $b(10, 0.36, 4 \text{ or fewer}) = 0.729$ or about 73%

6.69   a. $b(50, 0.42, \text{fewer than } 20) = 0.336$ or about 34%

b. $b(50, 0.42, \text{at most } 24) = 0.842$ or about 84%

c. $b(50, 0.42, \text{at least } 25) = 0.158$ or about 16%

6.71   a. $b(10, 0.90, 9) = 0.387$ or about 39%

b. $b(10, 0.90, 8 \text{ or fewer}) = 0.264$ or about 26%

c. $b(10, 0.90, \text{atleast } 9) = 0.736$ or about 74%

6.73   a. $b(40, 0.52, 20) = 0.121$ or about 12%

b. $b(40, 0.52, \text{fewer than } 20) = 0.340$ or about 34%

c. $b(40, 0.52, \text{at most } 20) = 0.461$ or about 46%

d. $b(40, 0.52, \text{between } 20 \text{ and } 23) = 0.463$ or about 46%

6.75   a. $b(50, 0.59, \text{at least } 25) = 0.924$ or about 92%

b. $b(50, 0.59, \text{ more than } 30) = 0.390$ or about 39%

c. $b(50, 0.59, \text{between } 30 \text{ and } 35) = 0.463$ or about 46%

d. $b(50, 0.41, \text{more than } 30) = 0.002$ or about 0.2%

6.77   a.   TT, TN, NT, NN

b. $P(TT) = (0.29)(0.29) = 0.0841$, $P(TN) = (0.29)(0.71) = 0.2059$, $P(NT) = (0.71)(0.29) = 0.2059$, $P(NN) = (0.71)(0.71) = 0.5041$

c. $P(NN) = 0.5041$ or about 50%

    d. $P(TT) = (0.29)(0.29) = 0.0841$ or about 8.4%

    e. $P(TN) + P(NT) = = 0.2059+ = 0.2059=$ or about 41.2%

6.79 a.   mean = (n)(p) = 200(0.53) = 106;

       Standard Deviation = $\sqrt{(n)(p)/(1-p)} = \sqrt{(200)(0.53)(1-0.53)} = 7.1$

    b.   Since 190 is more than two standard deviations from the mean, it would be surprising.

6.81 a.   mean = (n)(p) = 100(.55) = 55 people who are expected to pass.

    b.   Standard Deviation = $\sqrt{(n)(p)/(1-p)} = \sqrt{(100)(0.55)(1-0.55)} = 5$

    c. $55 - 2(5) = 45$, $55 + 2(5) = 65$; on about 95% of these days, the number of people passing will be as low as 45 and as high as 65.

    d.   Yes, 85 would be surprising because it is so far from the 95% interval given in part c.

Chapter Review Exercises

6.83 a.   discrete

    b.   continuous

    c.   continuous

    d.   discrete

6.85 95.2%, or about 95.2%

> **Normal Distribution**
> **Mean:20.5 Std. Dev.:0.9**
> **P(X≤22) = 0.95220965**

6.87 a.   0.7625, or about 76%   b. 98.6 °F

> **Normal Distribution**
> **Mean:20.5 Std. Dev.:0.9**
> **P(X≤19) = 0.04779035**

> **Normal Distribution**
> **Mean:98.1 Std. Dev.:0.7**
> **P(X≤98.6) = 0.76247474**

6.89 a.   One Standard Deviation below and above would be 225 - 15 = 210, 225 + 15 = 240, Using the Empirical Rule about 68%

    b.   1 - .68 = .32, .32/2 = 16%

    c.   A score of 255 is 30 pts or 2 standard deviation above the mean so approx 2.5%.

6.91 a.   0.106 or about 11%   b. Systolic blood pressures between 113 and 127

> **Normal Distribution**
> **Mean:120 Std. Dev.:8**
> **P(X≥130) = 0.10564977**

> **Normal Distribution**
> **Mean:120 Std. Dev.:8**
> **P(113≤X≤127) = 0.61842609**

    c.   0.394 or about 39%   d. Systolic blood pressure above 128

> **Normal Distribution**
> **Mean:120 Std. Dev.:8**
> **P(120≤X≤130) = 0.39435023**

> **Normal Distribution**
> **Mean:120 Std. Dev.:8**
> **P(X≥128.29147) = 0.15**

Note: The StatCrunch Binomial Calculator was used for the following calculations (6.93 – 6.98).

6.93 a.   *b*(200, 0.44, fewer than 80) = 0.113 or about 11%

    b.   *b*(200, 0.44, at least 90) = 0.414 or about 41%

c. b(200, 0.44, between 80 and 100) = 0.849 or about 85%

d. b(200, 0.44, at most 75) = 0.037 or about 4%

6.95 a. b(50, 0.46, more than 25) = 0.239 or about 24%

b. b(50, 0.46, at most 20) = 0.240 or about 24%

c. mean = (n)(p) = 50(0.46) = 23

Standard Deviation = $\sqrt{(n)(p)/(1-p)}$ = $\sqrt{(50)(0.46)(1-0.46)}$ = 3.52

d. It would be surprising because 10 is more than 3 standard deviations from the mean.

6.97 a. b(500, 0.67, more than 500) = about 100%

b. mean = (n)(p) = (500)(0.67) = 335

c. Standard Deviation = $\sqrt{(n)(p)/(1-p)}$ = $\sqrt{(500)(0.67)(1-0.67)}$ = 10.5 ; 450 is more than 3 standard deviations from the mean so it would be unusual to find that more than 450 out of 500 randomly selected Americans held this belief.

6.99 a. 0.0345 or 3.45%   b. $(0.0345)^2 = 0.0012$

```
Cumulative Distribution Function
Normal with mean = 7.5 and standard deviation = 1.1
   x   P( X <= x )
  5.5    0.0345182
```

c. 0.614   d. $200(0.0345) = 6.9$, or about 7

```
Cumulative Distribution Function
Binomial with n = 200 and p = 0.0345
x   P( X <= x )
7      0.613890
```

e. The standard deviation would be $\sqrt{200(0.0345)(0.9655)} = 2.58$.

f. Since $7 - 2(2.58) = 1.84$ and $7 + 2(2.58) = 12.16$, the interval would be about 2 to 12.

g. 45 would be surprising because it is so far from the interval of 2 to 12.

6.101 a. 19.7 inches   b. 20.5 inches

c. They are the same. The distribution is symmetric and so the mean is right in the middle.

```
Inverse Cumulative Distribution Function
Normal with mean = 20.5 and standard deviation = 0.9
P( X <= x )        x
      0.2    19.7425
      0.5    20.5000
```

Chapter 7: Survey Sampling and Inference

Section 7.1: Learning about the World through Surveys

Answers may vary slightly due to type of technology or rounding.

7.1 a. The population is adults in the United States. The sample consists of the 1021 adults surveyed.

  b. The parameter of interest is the percentage of all adults who support a ban on smoking in public places. The statistic is the 57% of the sample who supported such a ban.

7.3 a. The number 30,000 is a parameter because it describes all of Ross' paintings. The population is the collection of all of Ross' paintings.

  b. The number 10 percent is a statistic. It is based on a survey of PBS viewers who watched *The Joy of Painting*. The population is all PBS viewers who watched *The Joy of Painting*.

7.5 a. $\bar{x}$ is a statistic, and μ is a parameter.

  b. This is the mean of a sample, so it is $\bar{x}$.

7.7 μ is 218.8 pounds, the mean of the population, describes all professional NBA players, $\bar{x}$ is the 217.6 pounds, the mean of the sample.

7.9 a. 32470 is a sample mean, so the correct notation is $\bar{x}$.

  b. 28% is a sample proportion, so the correct notation is $\hat{p}$.

7.11 The sample is the 1207 survey participants. The population of interest is all Americans. The value 16% is a measure of the survey participants who are a sample of the population of interest. For this reason it is a statistic. We would use the symbol for a sample proportion: $\hat{p}$.

7.13 You want to test a sample of batteries because if you tested them all until they burned out, no usable batteries would be left.

7.15 First, all 10 cards are put in a bowl. Then one is drawn out and noted. When sampling with replacement, the card that is selected is replaced put back in the bowl, and a second draw is done. It is possible that the same student could be picked twice. When sampling without replacement, the first card is drawn and it is not put back in the bowl, so the second draw must be a different card.

7.17 The four friends chosen are 7, 3, 5, and 2.

  0 7 0 3 3   7 5 2 5 0   3 4 5 4 6

  7 5 2 9 8   3 3 8 9 3   6 4 4 8 7

7.19 Assign each student a pair of digits 00–29 (or 01–30). Read off pairs of digits from the random number table. It the number is within the range of assigned numbers, the student whose digits were called will be in the sample. Repeated numbers should be skipped and the selection will stop after the 10 students are selected.

7.21 One reason the district should be cautious because of the low survey response rate. The small percentage who chose to return the survey might be very different from the majority who did not return the survey.

7.23

	With persuasion	No persuasion
For Capital Punishment	$6 + 2 = 8$	$13 + 5 = 18$
Against Capital Punishment	$9 + 8 = 17$	$2 + 5 = 7$

  a. $(6+2)/(6+9+2+8) = 8/25 = 32\%$

  b. $(13+5)/(13+2+5+5) = 18/25 = 72\%$

  c. Yes, she spoke against it, and fewer who heard her statements against it (32%) supported capital punishment, compared with those who did not hear her persuasion (72%).

Section 7.2: Measuring the Quality of a Survey

7.25 a.   Sketches will vary. The sketch should show bullet holes consistently to the left of the target and close to each other. If the bullets go consistently to the left, then there is bias, not lack of precision.

b.   Sketches will vary. The sketch should show bullet holes that are all near the center of the target.

7.27 No, you cannot see bias from one numerical result. Bias is in the procedure, not the result. It is possible that the high mean happened by chance. "Statisticians evaluate the method used for a survey, not the outcome of a single survey."

7.29 a.   There are 17 odd numbers, so the proportion is 16/30, or about 53.3%, odd digits.

b.   16/30 is $\hat{p}$ (p-hat), the sample proportion.

c.   We would expect to get 15 odd numbers, so the error is $16/30 - 14/30 = 2/30$, or about 6.7%

7.31 a.   72%, the same as the population proportion.

b. $SE = \sqrt{\dfrac{p(1-p)}{n}} = \sqrt{\dfrac{0.72(1-0.72)}{100}} = .0449$ or 4.5%.

c.   We expect about 72% orange candies, give or take 4.5%.

d.   The standard error would decrease $SE = \sqrt{\dfrac{p(1-p)}{n}} = \sqrt{\dfrac{0.72(1-0.72)}{500}} = 0.02$ or 2%

7.33 The largest sample is the narrowest (the graph on the bottom), and the smallest sample is the widest (the graph on the top). Increasing the sample size makes the graph narrower.

7.35 The top dotplot has the largest standard error because it is widest, and the bottom dotplot has the smallest standard error because it is narrowest.

7.37 Graph A is for the fair coin, because it is centered at 0.50.

7.39 a.   No. We expect the sample proportion will be close to the population proportion (20%), but the sample proportions will vary from sample to sample with a standard error of

$SE = \sqrt{\dfrac{p(1-p)}{n}} = \sqrt{\dfrac{0.20(1-0.20)}{50}} = 0.057$ or 5.7%

b.   The sample proportion for a sample size of 500 will be closer to the population proportion because the standard error for samples of size 500 will be smaller than that of sample size 50.

$\left(SE = \sqrt{\dfrac{p(1-p)}{n}} = \sqrt{\dfrac{0.20(1-0.20)}{500}} = 0.018 \text{ or } 1.8\% \text{ compared to } 5.7\%. \right)$

Section 7.3: The Central Limit Theorem for Sample Proportions

7.41 Conditions for the CLT: The sample is random. The population of high school seniors is greater than 10(100) = 1000. 100(0.72) and 100(0.28) are both greater than 10.

The standard error is $SE = \sqrt{\dfrac{p(1-p)}{n}} = \sqrt{\dfrac{0.72(1-0.72)}{100}} = 0.045$.

The sampling distribution: N(0.72, 0.045). Use technology to find P(x ≥ 0.75) or find the z-score for 0.75): $(0.75 - 0.72)/0.045 = 0.67$ and find P(z ≥ 0.67).

The probability that more than 75% of the sample has a drivers license is 25.2%.

Normal Distribution
Mean:0.72 Std. Dev.:0.045
P(X≥0.75) = 0.25249254

7.43 a.   80%

b.   Sample is random; the population of Americans is greater than 10(1000) = 10,000; 1000(0.80) and 1000(0.20) are both greater than 10.

c.   standard error $= SE = \sqrt{\dfrac{p(1-p)}{n}} = \sqrt{\dfrac{0.80(1-0.80)}{1000}} = 0.013$

d. $0.80 \pm 2(0.013)$; between 0.774 and 0.826.

7.45 a.   60%

b.   sample is random; the population of Americans aged 18–29 is greater than 10(200) = 2000; 200(0.60) and 200(0.40) are both greater than 10.

c.   To determine if this is surprising find the $SE = \sqrt{\dfrac{p(1-p)}{n}} = \sqrt{\dfrac{0.60(1-0.60)}{200}} = 0.035$, 125/200 = 0.625; z
$= (0.625 - 0.60) / 0.035 = 0.71$; No, this not surprising because it is less than 1 standard deviation from the mean.

d. $z = (0.74 - 0.60) / 0.035 = 4$; Yes, this would be surprising because it is more than 3 standard deviations from the mean.

7.47 a.   We expect the percentage of Americans who feel this way to be about 0.59. Because 0.55 is less than this and because the sampling distribution is approximately Normal, there would be a greater than 50% chance of seeing a sample proportion greater than 55%.

b.   The sampling distribution is Normal, the mean is 0.59, and the standard error is

$SE = \sqrt{\dfrac{p(1-p)}{n}} = \sqrt{\dfrac{0.59(1-0.59)}{200}} = 0.035$, z $= (0.55 - 0.59) / 0.035 = -1.14$

The area to the right of a z-score of -1.14 is 0.873.
Normal Distribution
Mean:0 Std. Dev.:1
P(X≥-1.14) = 0.87285685

7.49 The sample is random; the population of Americans is greater than 10(120) = 1200; 120(0.45) and 120(0.55) are both greater than 10 so the conditions for the Central Limit Theorem are met.

The Sampling Distribution is approximately Normal with a mean = 0.45 and

a standard error $= SE = \sqrt{\dfrac{p(1-p)}{n}} = \sqrt{\dfrac{0.45(1-0.45)}{120}} = 0.045$; $z = (0.50 - 0.45) / 0.045 = 1.11$.

The area to the right of a z-score of 1.11 is 0.133.
Normal Distribution
Mean:0 Std. Dev.:1
P(X≥1.11) = 0.13349951

7.51 This probability cannot be calculated because the sample is too small to satisfy the conditions of the Central Limit Theorem. $np = 100(0.005) = 0.5 < 10$.

Section 7.4: Estimating the Population Proportion with Confidence Intervals

7.53 a.   64.0% and 70.0%.

   b.   Yes, because the interval only includes values that are greater than 50%.

7.55 a. $\hat{p} = \dfrac{617}{1028} = 0.60,$ or 60.0%.

   b.   1: *Random and Independent*: Gallup Polls are random samples, and the people selected are independent.

   2: *Large Sample*: Because we do not know p, we will use $\hat{p}$ for these calculations;

   $n\hat{p} = 1028(0.60) = 617$ and $n(1 - \hat{p}) = 1028(1 - 0.60) = 411$, and both of these are more than 10.

   3: *Big Population*: The population of the United States is more than $10(1028) = 10,280$.

   c.   I am 95% confident the population proportion of adults living in the United States who feel the laws covering the sale of firearms should be more strict is between 0.570 and 0.630;

   $$\hat{p} \pm z^* \sqrt{\dfrac{\hat{p}(1 - \hat{p})}{n}} = 0.60 \pm 1.96 \sqrt{\dfrac{0.60(0.40)}{1028}} = 0.60 \pm 0.030$$

   d.   Yes, because the confidence interval only includes values greater than 50%.

7.57 a.   I am 95% confident that the population percentage of voters supporting Candidate X is between 53% and 57%.

   b.   There is no evidence that he could lose, because the interval is entirely above 50%.

   c.   A sample from New York City would not be representative of the entire country and would be worthless in this context.

7.59 a. $SE_{est} = \sqrt{\dfrac{p(1 - p)}{n}} = \sqrt{\dfrac{0.26(1 - 0.26)}{1207}} = 0.013.$

   b. $\hat{p} \pm z^* \sqrt{\dfrac{\hat{p}(1 - \hat{p})}{n}} = 0.26 \pm 1.96(0.013) = 0.26 \pm 0.025$; between $(0.235, 0.285)$

   c. $m = z^* (SE_{est}) = 1.96(0.013) = 0.025,$ or 2.5%

   d.   No because 0.247 is contained in the confidence interval. It is possible the proportion is still the same.

7.61 a.   (0.602, 0.679); We are 99% confident that the proportion of all U.S. adults who believe marijuana should be legalized is between 60.2% and 66.79%.

   99% confidence interval results:

Proportion	Count	Total	Sample Prop.	Std. Err.	L. Limit	U. Limit
p	658	1028	0.64007782	0.014970081	0.60151745	0.67863819

   b.   (0.611, 0.669); We are 95% confidence that the proportion of all U.S. adults who believe marijuana should be legalized is between 61.1% and 66.9%.

   95% confidence interval results:

Proportion	Count	Total	Sample Prop.	Std. Err.	L. Limit	U. Limit
p	658	1028	0.64007782	0.014970081	0.610737	0.66941864

   c.   For part a. $m = z^* (SE_{est}) = 2.58(0.0150) = 0.0387$, for part b. $m = z^* (SE_{est}) = 1.96(0.0150) = 0.0294,$

   d.   A 90% confidence interval will be narrower than a 95% or 99%. As confidence level decreases, the width of the interval also decreases.

7.63   a.   0.74(1000) = 740 people

b.   1000(0.74) = 740 and 1000(1 - 0.74) = 260. Since both of these values are greater than 10 the Central Limit Theorem can be applied.

c.   (0.713, 0.767)

95% confidence interval results:

Proportion	Count	Total	Sample Prop.	Std. Err.	L. Limit	U. Limit
p	740	1000	0.74	0.013870833	0.71281367	0.76718633

d.   The width of the 95% confidence interval is 0.767 – 0.713 = 0.054 or about 5.4%.

e.   (0.726, 0.754); the width of the interval is 0.754 – 0.726 = 0.028.

95% confidence interval results:

Proportion	Count	Total	Sample Prop.	Std. Err.	L. Limit	U. Limit
p	2960	4000	0.74	0.0069354164	0.72640683	0.75359317

f.   The width of the interval decreased when the sample size was multiplied by 4.

7.65   a.   (0.37, 0.43); we are 95% confident that the percentage of American adults who were very likely to watch some of the Winter Olympic coverage on television is between 37% and 43%.

b.   We would expect at least 95 of them to include the true population proportion.

c.   This interpretation is incorrect because a confidence interval is about a population not a sample.

7.67   a.   34226731/68531917, or 49.9%, of the voters voted for Kennedy.

b.   No, you should not find a confidence interval. We have the population proportion. We need a confidence interval only when we have a sample statistic such as a sample proportion and want to generalize about the population from which the sample was drawn.

7.69   a.   438/1019 = 0.43

b.   (0.40, 0.46)

95% confidence interval results:

Proportion	Count	Total	Sample Prop.	Std. Err.	L. Limit	U. Limit
p	438	1019	0.42983317	0.015508288	0.39943748	0.46022885

c.   A 90% confidence interval would be narrower than the 95% interval because a 90% confidence interval includes less of the sampling distribution.

d.   (0.404, 0.455). The 90% interval is narrower than the 95% interval. As confidence level decreases, the width of the interval also decreases.

90% confidence interval results:

Proportion	Count	Total	Sample Prop.	Std. Err.	L. Limit	U. Limit
p	438	1019	0.42983317	0.015508288	0.40432431	0.45534203

7.71   a. $1/(0.06)^2$ = 278 Americans

b. $1/(0.04)^2$ = 625 Americans

c.   As the margin of error is decreased, the required sample size increases.

Section 7.5: Comparing Two Population Proportions with Confidence

7.73   The negative value in the interval indicates the first population proportion (proportion happy in 2016) is less than the second population proportion (proportion happy in 2017). We are 95% confidence the difference in the population proportions is between (-0.06 and 0.02). The confidence interval contains 0. This tells us that the proportion of people who reported being happy in 2016 and 2017 were not significantly different.

7.75   a.   No, because these are sample proportions, not population proportions.

b.   (1) Random and independent samples: Gallup takes random samples and the samples are independent;

(2) Large samples: $\hat{p} = (582 + 828) / (1200 + 1200) = 0.565$; $1200(0.565) = 678$; $1200(1 - 0.565) = 522$; both these values are greater than 10;

(3) Big populations: there are more than $10(1200) = 12,000$ Democratic voters.

c.   (-0.288, -0.212)

95% confidence interval results:

Difference	Count1	Total1	Count2	Total2	Sample Diff.	Std. Err.	L. Limit	U. Limit
$p_1 - p_2$	528	1200	828	1200	-0.25	0.019585284	-0.28838645	-0.21161355

d.   The interval does not contain 0. Since both numbers are negative, the $p_2 > p_1$. A greater percentage of Democratic voters thought the FBI did a good or excellent job in 2017 than in 2003. We are 95% confident the difference in the population proportions is between 0.212 and 0.288.

7.77   1. *Random and Independent:* Although we do not have random samples, we have random assignment to groups.

2. *Large Samples:* $n \hat{p}_1 = 57(0.65) = 37$, $n_1(1 - \hat{p}_1) = 57(0.35) = 20$, $n \hat{p}_2 = 64(0.45) = 29$, $n_2(1\ \hat{p}_2) = 64(0.55) = 35$; all four numbers are more than 10.

3. *Big Populations:* There are far more than $10(57)$ African-American boys in the United States who go to preschool and more than $10(64)$ African-American boys who do not go to preschool.

4. *Independent Samples*: The boys in the Preschool group are not related to the boys in the No Preschool control group.

The confidence interval is (0.022, 0.370).

95% confidence interval results:

Difference	Count1	Total1	Count2	Total2	Sample Diff.	Std. Err.	L. Limit	U. Limit
$p_1 - p_2$	37	57	29	64	0.19599781	0.088700333	0.022148349	0.36984727

7.79   a.   Fish oil: 58/346 = 0.168 or 16.8%; Placebo: 83/349 = 0.238 or 23.8%;

b.   (1) Random and independent: Although we do not have random samples, we have random assignment to groups. (2) Large samples: $\hat{p} = (58 + 83) / (346 + 349) = 141 / 695 = 0.203$ or 20.3%; %; $346(0.203) = 70$; $346(1 - 0.203) = 339$; $349(0.203) = 71$; $349(1 - 0.203) = 278$; all of these values are greater than 10. (3) Big population: The population of children in Denmark is larger than $10(695) = 6950$.

c.   (-0.130, -0.011); since the interval does not contain 0, there is a difference in the population proportions. Since both numbers are negative, $p_2 > p_1$. A smaller proportion of children whose mothers took fish oil during pregnancy developed asthma. We are 95% confident the difference in the population proportions is between 0.013 and 0.011.

95% confidence interval results:

Difference	Count1	Total1	Count2	Total2	Sample Diff.	Std. Err.	L. Limit	U. Limit
$p_1 - p_2$	58	346	83	349	-0.070192292	0.030375062	-0.12972632	-0.010658264

7.81   a.   585/936, or 62.5%, of men and 452/607, or 74.5%, of women used turn signals.

b. (–0.166, –0.073); *Large Samples*, *Random Samples*, *Independent Samples*, *Independent Observations*, and *Big Populations* are given. I am 95% confident that the population percentage of men using turn signals minus the population percentage of women using turn signals is between −16.6% and −7.3%. The interval does not capture 0, so we are confident that the percentages are different. This shows that women are more likely than men to use turn signals.

$$\left(\hat{p}_2-\hat{p}_1\right)\pm z^*\sqrt{\frac{\hat{p}_1\left(1-\hat{p}_1\right)}{n_1}+\frac{\hat{p}_2\left(1-\hat{p}_2\right)}{n_2}}=\left(\frac{585}{936}-\frac{452}{607}\right)\pm1.96\sqrt{\frac{\left(\frac{585}{936}\right)\left(\frac{351}{936}\right)}{936}+\frac{\left(\frac{452}{607}\right)\left(\frac{155}{607}\right)}{607}}$$

$$=-0.1196\pm0.0465$$

95% confidence interval results:

Difference	Count1	Total1	Count2	Total2	Sample Diff.	Std. Err.	L. Limit	U. Limit
$p_1 - p_2$	585	936	452	607	-0.1196458	0.023741525	-0.16617833	-0.073113265

c. 59/94, or 62.8%, of the men and 45/61, or 73.8%, of the women used turn signals. (−0.2575, 0.03739); I am 95% confident that the population percentage of men using turn signals minus the population percentage of women using turn signals is between –25.8% and 3.7%. The interval captures 0, showing it is plausible that the percentages are the same in the population.

$$\left(\hat{p}_2-\hat{p}_1\right)\pm z^*\sqrt{\frac{\hat{p}_1\left(1-\hat{p}_1\right)}{n_1}+\frac{\hat{p}_2\left(1-\hat{p}_2\right)}{n_2}}=\left(\frac{59}{94}-\frac{45}{61}\right)\pm1.96\sqrt{\frac{\left(\frac{59}{94}\right)\left(\frac{35}{94}\right)}{94}+\frac{\left(\frac{45}{61}\right)\left(\frac{16}{61}\right)}{61}}$$

$$=-0.1100\pm0.1474$$

95% confidence interval results:

Difference	Count1	Total1	Count2	Total2	Sample Diff.	Std. Err.	L. Limit	U. Limit
$p_1 - p_2$	59	94	45	61	-0.11004534	0.075221501	-0.25747678	0.037386088

With more data (part b), we had a narrower margin of error and thus a more precise estimate of the true difference in proportions. We could be confident that the percentages were different in the population. In part d, the very wide interval did not allow us to make this call.

7.83 a. No, the rate of miscarriages was higher for the unexposed women, which is the opposite of what was feared.

b. There was not a random, and sample, and there was not random assignment.

Chapter Review Exercises

7.85 a. 0.45

b. $SE_{est}=\sqrt{\frac{\hat{p}\left(1-\hat{p}\right)}{n}}=\sqrt{\frac{0.45\left(0.55\right)}{500}}=0.022$

c. 45%, give or take 2.2%

d. $z=\left(0.55-0.45\right)/0.022=4.55$; Yes, this would be surprising because 0.55 is more than 4 standard errors from the mean.

e. Decreasing the sample size would increase the standard error.

7.87 a. $0.52\pm0.04=(0.48, 0.56)$

b. If the sample size were larger, the standard error would be smaller. The resulting interval would be narrower than the one in part a.

c. If the confidence level were 90% rather than 95%, the interval would be narrower than the one in part a.

d. The population size does not have any effect on the width of the interval.

7.89   The sample proportion must be $(.40 + .48) / 2 = 44\%$ because the interval is symmetric around the sample proportion, which is in the middle.

7.91   The margin of error must be 4%. From the sample proportion to find the upper boundary you go up one margin of error and to find the lower boundary you go down one margin of error. Therefore, the boundaries are separated by two margins of error, and half of 8% is 4%; $(48\% - 40\%) / 2 = 4\%$.

7.93   *Large Sample*: $np = 200(0.29) = 58 > 10$ and $n(1 - p) = 200(0.71) = 142 > 10$. *Random Sample, Independent Observations*, and *Big Population* are met.

$$p = 0.29 \text{ and } SE = \sqrt{\frac{p(1-p)}{n}} = \sqrt{\frac{0.29(0.71)}{200}} = 0.03209.$$

For 50% or higher, $z = \dfrac{\hat{p} - p}{SE} = \dfrac{0.50 - 0.29}{0.03209} = 6.54.$ The area of the normal curve to the right of a z-value of 6.54 is less than 0.001. Therefore, the probability that a sample of 200 will contain 50% or more dreaming in color is less than 0.001.

Cumulative Distribution Function
```
Normal with mean = 0 and standard deviation = 1
  x   P( X <= x )
6.54            1.00000
```
Cumulative Distribution Function
```
Normal with mean = 0.29 and standard deviation = 0.03209
  x   P( X <= x )
0.5            1.00000
```

7.95   The confidence interval for Clinton would be $0.45 \pm 0.052 = (0.398, 0.502)$. Since the interval contains values that are less than 50%, we cannot predict with confidence that she would win in an election with Sanders.

7.97   a.   2008: $623/1022 = 0.610$; 2017: $460/1022 = 0.450$;

   b. $(0.117, 0.202)$; Since the interval does not contain 0, there is a difference in the population proportions. Since both numbers are positive, $p_1 > p_2$. A greater proportion of people trusted the executive branch in 2008 than in 2017. We are 95% confident the difference in the population proportions is between 0.117 and 0.202.

$$(\hat{p}_2 - \hat{p}_1) \pm z^* \sqrt{\frac{\hat{p}_1(1 - \hat{p}_1)}{n_1} + \frac{\hat{p}_2(1 - \hat{p}_2)}{n_2}} = \left(\frac{623}{1022} - \frac{460}{1022}\right) \pm 1.96 \sqrt{\frac{\left(\frac{623}{1022}\right)\left(\frac{399}{1022}\right)}{1022} + \frac{\left(\frac{460}{1022}\right)\left(\frac{562}{1022}\right)}{1022}}$$

$$= 0.160 \pm 0.043$$

95% confidence interval results:

Difference	Count1	Total1	Count2	Total2	Sample Diff.	Std. Err.	L. Limit	U. Limit
$p_1 - p_2$	624	1022	460	1022	0.16046967	0.021790782	0.11776052	0.20317881

7.99   We would need a sample size of 1111 or 1112 to get a margin of error of 3 percentage points, since

$$n = \frac{1}{m^2} = \frac{1}{(0.03)^2} = 1111.11.$$

Thus, we would need a sample size of 1111 or 1112 to get a margin of error of no more than 3 percentage points.

7.101   Marco took a convenience sample. The students may not be representative of the voting population, so the proposition may not pass.

7.103   No, the people you met would not be a random sample but a convenience sample.

7.105   The small mean might have occurred by chance.

7.107 $m = 2\sqrt{\dfrac{\hat{p}(1 - \hat{p})}{n}} = \sqrt{\dfrac{0.50(0.50)}{n}}$, simplifying we have $m = 2\sqrt{\dfrac{0.25}{n}}.$

Chapter 8: Hypothesis Testing for Population Proportions

Answers may vary slightly due to type of technology or rounding.

Section 8.1: The Essential Ingredients of Hypothesis Testing

8.1 The null hypothesis is always a statement about a <u>population parameter</u>.

8.3 H_0: The proportion of American adults who are vegetarian is 0.033, H_0: $p = 0.033$; H_a: The proportion of American adults who are vegetarian is greater than 0.033, H_a: $p > 0.033$.

8.5 a. i

   b. ii

   c. i

8.7 ii.

8.9 The null hypothesis is written correctly; the alternative hypothesis should be $p > 0.15$, since the promotion is successful if the proportion of customers ordering a soda with their meal has increased.

8.11 The probability of saying the current flu vaccine is less than 73% effective against the flu virus when, in fact, it isn't is 0.01.

8.13 a. H_0: $p = 0.37$, H_a: $p \neq 0.37$

   b. $z = 0.6807$

8.15 a. $\hat{p} = \dfrac{11}{150} = 0.073$

   b. $p_0 = 0.033$

   c. $z = \dfrac{\hat{p} - p_0}{\sqrt{\dfrac{p_0(1 - p_0)}{n}}} = \dfrac{0.073 - 0.033}{\sqrt{\dfrac{0.033(1 - .033)}{150}}} = \dfrac{0.04}{0.0145856} = 2.74$

   The value of the test statistic tells us that the observed proportion was 2.74 standard errors above the null hypothesis value of 0.033.

8.17 a. You would expect about $20(0.5) = 10$ right out of 20 if the person is guessing.

   b. The smaller p-value will come from person B. The p-value measures how unusual an event is, assuming the null hypothesis is true. Getting 18 right out of 20 is more unusual than getting 13 right out of 20 when you are expecting 10 right under the null hypothesis. The larger difference in proportions, $\hat{p} - p_0$, results in a smaller p-value.

8.19 The probability of finding 11 or more vegetarians in a random sample of 150 American adults assuming the population proportion is 0.033. Since the probability 0.0028 is very small, the nutritionist should not believe the null hypothesis is true.

8.21 a. $\hat{p} = \dfrac{11}{70} = 0.157$

   b. H_0: $p = 0.17$, H_a: $p < 0.17$

   c. $z = \dfrac{\hat{p} - p_0}{\sqrt{\dfrac{p_0(1 - p_0)}{n}}} = \dfrac{0.157 - 0.17}{\sqrt{\dfrac{0.17(.83)}{70}}} = \dfrac{-0.013}{0.0449} = -0.29$

The value of the test statistic tells us that the observed proportion was 0.29 standard errors below the null hypothesis value of 0.17.

d.   The probability of 11or fewer pneumonia patients out of a sample of 70 patients being readmitted to the hospital is 0.39, assuming the population proportion is 0.17. Since the probability is not close to zero, we do not doubt the null hypothesis is true.

Section 8.2: Hypothesis Testing in Four Steps

8.23   Random Sample: Given

Large Sample: $np_0 = 113(0.29) = 32.77 > 10$ and $n(1 - p_0) = 113(0.71) = 80.23 > 10$

Large Population: There are more than $10 \times 113 = 1130$ people in the population of dreamers.

Independence: May be assumed.

So the conditions are met.

8.25   Figure A correctly matches the alternative hypothesis p > 0.30 because it is one-tailed. The p-value is 0.08. If the population proportion of young Americans who would be comfortable riding in a self-driving car is 30%, there is about an 8% chance of getting 152 or more out of 461 randomly selected adults from this age group who feel this way.

8.27   Figure B is correct because the alternative hypothesis should be one-sided, since the person should get better than half right if she or he can tell the difference.

8.29   Step 1: H$_0$: $p = 0.73$, H$_a$: $p > 0.73$.

Step 2: One-proportion z-test,

Random Sample: Given Independence: Assumed.

Large Sample: $np_0 = 200(0.73) = 146 > 10$ and $n(1 - p_0) = 200(0.27) = 54 > 10$

Large Population: There are more than $10 \times 200 = 2000$ people in the population of the United States.

Entries: p$_0$: 0.73, x: 160, n: 200, one sided, >

8.31   Step 3: Significance level = 0.05; z = 2.23, p-value = 0.01.

Step 4: Reject H$_0$. The proportion of Americans who report working out one or more times each week has increased.

8.33   In Figure (A), the shaded area could be a p-value because it includes tail areas only; it would be for a two-sided alternative because both tails are shaded. In Figure (B) the shaded area would not be a p-value because it is the area between two z-values.

8.35   a.   Step 1: H$_0$: $p = 0.47$, H$_a$: $p > 0.47$, where p is the population proportion favoring stricter gun control.

Step 2: One-proportion z-test,

Random Sample: Given Independence: Assumed.

Large Sample: $np_0 = 3635(0.47) = 1708 > 10$ and $n(1 - p_0) = 3635(0.53) = 1927 > 10$

Large Population: There are more than $10 \times 3635 = 36,350$ Facebook users

Step 3: Significance level = 0.05. z = 25.37, p-value approximately 0.

$$\hat{p} = \frac{2472}{3635} = .68 \quad z = \frac{\hat{p} - p_0}{\sqrt{\dfrac{p_0(1 - p_0)}{n}}} = \frac{0.68 - 0.47}{\sqrt{\dfrac{0.47(0.53)}{3635}}} = 25.37$$

Step 4: Reject H_0. The proportion of Facebook users who get their world news on Facebook has increased since 2013.

b. (0.667, 0.693). The interval supports the hypothesis test conclusion because it only includes values that are greater than 0.47, supporting a conclusion that the population proportion has increased.

90% confidence interval results:

Proportion	Count	Total	Sample Prop.	Std. Err.	L. Limit	U. Limit
p	2472	3635	0.68005502	0.0077367305	0.66732923	0.69278081

8.37 a. $\hat{p} = \dfrac{692}{1018} = .68$

b. Step 1: H_0: p = 0.57, H_a: p ≠ 0.57.

Step 2: 1-proportion z-test. 0.68(1018) = 692 >10; 0.32(1018) = 326 > 10;

Random sample is given and assumed independent; large population: # Americans > 10(1018).

Step 3: Significance level = 0.01; z = 7.07, p-value approximately 0.

$$z = \frac{\hat{p} - p_0}{\sqrt{\dfrac{p_0(1-p_0)}{n}}} = \frac{0.68 - 0.57}{\sqrt{\dfrac{0.57(0.43)}{1019}}} = 7.07$$

Step 4: Reject H_0.

c. Choice ii is correct: In 2017, the percentage of Americans who believe global warming is caused by human activities have changed from the historical level of 0.57.

8.39 Step 1: H_0: p = 0.20, H_a: p ≠ 0.20, where p is the population proportion of dangerous fish.

Step 2: One-proportion z-test, 0.2(250) = 50 > 10 and 0.8(250) = 200 > 10, population large, assume a random and independent sample.

Step 3: α= 0.05. z = 1.58, p-value = 0.114.

$$\hat{p} = \frac{60}{250} = .24 \quad z = \frac{\hat{p} - p_0}{\sqrt{\dfrac{p_0(1-p_0)}{n}}} = \frac{0.24 - 0.20}{\sqrt{\dfrac{0.20(0.80)}{250}}} = 1.58$$

```
Test and CI for One Proportion
Test of p = 0.2 vs p not = 0.2
Sample   X     N    Sample p          95% CI          Z-Value    P-Value
1        60    250  0.240000   (0.187059, 0.292941)    1.58      0.114
Using the normal approximation.
```

Step 4: Do not reject H_0, 0.114 > 0.05. We are not saying the percentage is 20%. We are only saying that we cannot reject 20%. (We might have been able to reject the value of 20% if we had had a larger sample.)

8.41 Step 1: H_0: p = 0.09, H_a: p ≠ 0.09, where p is the population proportion of t's in the English language.

Step 2: One-proportion z-test, $\alpha = 0.10$

Random Sample: Given

Large Sample: $np_0 = 600(0.09) = 54 > 10$ and $n(1-p_0) = 600(0.91) = 546 > 10$

Large Population: There are more than $10 \times 600 = 6000$ letters in the population of letters in all the words in the English language.

Independence: Assumed

Step 3: $\hat{p} = \dfrac{48}{600} = 0.08$, $z = \dfrac{\hat{p} - p_0}{\sqrt{\dfrac{p_0(1 - p_0)}{n}}} = \dfrac{0.08 - 0.09}{\sqrt{\dfrac{0.09(0.91)}{600}}} = -0.86$, p-value $= 0.392$.

```
Test and CI for One Proportion
Test of p = 0.09 vs p not = 0.09
Sample   X    N  Sample p          90% CI          Z-Value  P-Value
1        48  600  0.080000  (0.061782, 0.098218)    -0.86    0.392
Using the normal approximation.
```

Step 4: Do not reject H_0. We cannot reject 9% as the current proportion of t's because 0.392 is more than 0.10.

Section 8.3: Hypothesis Tests in Detail

8.43 iv. $z = 3.00$; It is farthest from 0 and therefore has the smallest tail area.

8.45 The first kind of error is concluding more than 50% of eligible voters from this age group voted in the 2016 election when, in fact, the percentage who voted was not more than 50%. The second type of error is concluding that the percentage from this age group who voted isn't more than 50% when, in fact, it is.

8.47 The first type of error is having the innocent person suffer (convicting an innocent person). The second type of error is "ten guilty persons escape" (letting guilty persons go free).

8.49 With a significance level of 0.01 and a p-value of 0.02, we would not reject H_0; however, we have not proved the null hypothesis is true. The conclusion for this hypothesis test is that, using a 1% significance level, there is not enough evidence to conclude the proportion of Americans who would pick invisibility as their superpower has increased.

8.51 Choose hypothesis testing and the one-proportion z-test, because he only wants to know whether or not it will pass; he is not interested in knowing the proportion who will vote for it.

Step 1: H_0: $p = 0.50$, H_a: $p > 0.50$, where p is the population proportion supporting the proposition.

Step 2: One-proportion z-test, $\alpha = 0.05$

Random Sample: Given

Large Sample: $np_0 = 1000(0.50) = 500 > 10$ and $n(1 - p_0) = 1000(0.50) = 500 > 10$

Large Population: There are more than $10 \times 1000 = 10,000$ people in the population of the United States.

Independence: Assumed

Step 3: $\hat{p} = \dfrac{580}{1000} = 0.58$, $z = \dfrac{\hat{p} - p_0}{\sqrt{\dfrac{p_0(1 - p_0)}{n}}} = \dfrac{0.58 - 0.50}{\sqrt{\dfrac{0.50(0.50)}{1000}}} = 5.06$, p-value < 0.001.

```
Test and CI for One Proportion
Test of p = 0.5 vs p > 0.5
                            95% Lower
Sample    X     N  Sample p   Bound   Z-Value  P-Value
1        580  1000  0.580000  0.554328   5.06    0.000
Using the normal approximation.
```

Step 4: Reject H_0. The proposition is likely to pass.

8.53 Interpretation iii is the correct choice.

8.55 No, we don't use "prove" because we cannot be 100% sure of conclusions based on chance processes.

8.57 It is a null hypothesis.

8.59   Interpretations b and d are valid. Interpretations a and c are both "accepting" the null hypothesis claim, which is an incorrect way of expressing the outcome.

Section 8.4: Comparing Proportions from Two Populations

8.61   Far apart. Assuming the standard errors are the same, the farther apart the two proportions are, the larger the absolute value of the numerator of z, and therefore the larger the absolute value of z and the smaller the p-value.

8.63   a.   ritonavir-boosted darunavir: $306/382 = 0.801$, dorovirine: $321/382 = 0.840$

b.   Step 1: H_0: $p_{rbd} = p_{dorovine}$, H_a: $p_{rbd} \neq p_{dorovine}$

Step 2: Two-proportion z-test. Because the two sample sizes are equal ($n_1 = n_2$) the numbers below are the same for both samples. $\hat{p}$ = 306+321/382+382 = 0.82; $n\,\hat{p}$ = 382(0.82) = 313 > 10,

$n(1 - \hat{p}) = 382(0.18) = 69 > 10$.

Random assignment and independence within and between samples.

Step 3: Significance level = 0.01. z = -1.41, p-value = 0.16.

$$z = \frac{(\hat{p}_1 - \hat{p}_2 - 0)}{\sqrt{\hat{p}(1-\hat{p})\left(\dfrac{1}{n_1}+\dfrac{1}{n_2}\right)}} = \frac{\left(\dfrac{306}{382} - \dfrac{321}{382} - 0\right)}{\sqrt{\dfrac{627}{764}\left(1-\dfrac{627}{764}\right)\left(\dfrac{1}{382}+\dfrac{1}{382}\right)}} = -1.41$$

Step 4: Do not reject H_0. There is no evidence that there is a difference in the proportion of patients who achieve a positive outcome between the two treatments. Based on this study we have no evidence that dorovirine might be a more effective treatment option for HIV-1 than ritonavir-boosted darunavir.

p_1 : proportion of successes for population 1
p_2 : proportion of successes for population 2
p_1 - p_2 : Difference in proportions
H_0 : p_1 - p_2 = 0
H_A : p_1 - p_2 ≠ 0

Hypothesis test results:

Difference	Count1	Total1	Count2	Total2	Sample Diff.	Std. Err.	Z-Stat	P-value
p_1 - p_2	306	382	321	382	-0.039267016	0.027757734	-1.4146333	0.1572

8.65   a.   2015: $1201/1906 = 0.63$, 2018: $1341/2002 = 0.67$

b. $\hat{p}$ = 1201 + 1341/1906 + 2002 = 0.65

c. $$z = \frac{(\hat{p}_1 - \hat{p}_2 - 0)}{\sqrt{\hat{p}(1-\hat{p})\left(\dfrac{1}{n_1}+\dfrac{1}{n_2}\right)}} = \frac{\left(\dfrac{1201}{1906} - \dfrac{1341}{2002} - 0\right)}{\sqrt{\dfrac{2542}{3908}\left(1-\dfrac{2542}{3908}\right)\left(\dfrac{1}{1906}+\dfrac{1}{2002}\right)}} = -2.60$$

p_1 : proportion of successes for population 1
p_2 : proportion of successes for population 2
p_1 - p_2 : Difference in proportions
H_0 : p_1 - p_2 = 0
H_A : p_1 - p_2 < 0

Hypothesis test results:

Difference	Count1	Total1	Count2	Total2	Sample Diff.	Std. Err.	Z-Stat	P-value
p_1 - p_2	1201	1906	1341	2002	-0.039714745	0.015259572	-2.6026119	0.0046

Reject H_0,, There is evidence to support the proportion who read print books has increased.

8.67 a. Step 1: H_0: $p_{2016} = p_{2017}$, H_a: $p_{2016} \neq p_{2017}$.

Step 2: Two-proportion z-test. $\hat{p} = 2489+1808/3072+2015 = 0.845$; $n_1 \hat{p}_1 = 3072(0.845)$, $n_1(1 - \hat{p}_1) = 3072(0.155)$, $n_2 \hat{p}_2 = 2014(0.845)$, $n_2(1 - \hat{p}_2) = 2014(0.155)$; all expected counts are greater than 10. Samples are random and assumed independent.

Step 3: Significance level: 0.05. $z = -8.43$; *p-value* is approximately 0.

$$z = \frac{(\hat{p}_1 - \hat{p}_2 - 0)}{\sqrt{\hat{p}(1-\hat{p})\left(\dfrac{1}{n_1} + \dfrac{1}{n_2}\right)}} = \frac{\left(\dfrac{2489}{3072} - \dfrac{1808}{2015} - 0\right)}{\sqrt{\dfrac{4297}{5089}\left(1 - \dfrac{4297}{5089}\right)\left(\dfrac{1}{13072} + \dfrac{1}{2015}\right)}} = -8.43$$

p_1 : proportion of successes for population 1
p_2 : proportion of successes for population 2
$p_1 - p_2$: Difference in proportions
H_0 : $p_1 - p_2 = 0$
H_A : $p_1 - p_2 \neq 0$

Hypothesis test results:

Difference	Count1	Total1	Count2	Total2	Sample Diff.	Std. Err.	Z-Stat	P-value
$p_1 - p_2$	2489	3072	1808	2014	-0.087494634	0.010379888	-8.4292465	<0.0001

Step 4: Reject H_0. The proportion of college students who believe that freedom of the press is secure or very secure in the country changed from 2016.

b. The 95% confidence interval for $p_{2016} - p_{2017} = (-0.107, -0.068)$. Since the interval does not include 0, the population proportions are significantly different. The proportion of college students who believe that freedom of the press is secure declined from 2016 and the difference is between 6.8% and 10.7%.

p_1 : proportion of successes for population 1
p_2 : proportion of successes for population 2
$p_1 - p_2$: Difference in proportions

95% confidence interval results:

Difference	Count1	Total1	Count2	Total2	Sample Diff.	Std. Err.	L. Limit	U. Limit
$p_1 - p_2$	2489	3072	1808	2014	-0.087494634	0.0097798169	-0.10666272	-0.068326545

8.69 Step 1: H_0: $p_{2016} = p_{2018}$, H_a: $p_{2016} > p_{2018}$.

Step 2: Two-proportion z-test. There were 983 sampled in 2016 and 993 sampled in 2018. $\hat{p} = 543+461/983+983 = 0.508$; $n_1 \hat{p}_1 = 983(0.508)$, $n_1(1 - \hat{p}_1) = 983(0.492)$, $n_2 \hat{p}_2 = 983(0.508)$, $n_2(1 - \hat{p}_2) = 983(0.492)$; all expected counts are greater than 10. Samples are random and assumed independent.

Step 3: Significance level: 0.05. $z = 3.70$, p-value is approximately 0.

$$z = \frac{(\hat{p}_1 - \hat{p}_2 - 0)}{\sqrt{\hat{p}(1-\hat{p})\left(\dfrac{1}{n_1} + \dfrac{1}{n_2}\right)}} = \frac{\left(\dfrac{543}{983} - \dfrac{461}{983} - 0\right)}{\sqrt{\dfrac{1004}{1966}\left(1 - \dfrac{1004}{1966}\right)\left(\dfrac{1}{983} + \dfrac{1}{983}\right)}} = 3.70$$

p_1 : proportion of successes for population 1
p_2 : proportion of successes for population 2
$p_1 - p_2$: Difference in proportions
H_0 : $p_1 - p_2 = 0$
H_A : $p_1 - p_2 > 0$

Hypothesis test results:

Difference	Count1	Total1	Count2	Total2	Sample Diff.	Std. Err.	Z-Stat	P-value
$p_1 - p_2$	543	983	461	983	0.083418108	0.022548057	3.6995697	0.0001

Step 4: Reject H₀. The proportion of Americans who are satisfied with the quality of the environment has declined.

Chapter Review Exercises

8.71 a.   One-proportion z-test. The population is voters in California.

b.   Two-proportion z-test. Populations are residents of coastal states and residents of non-coastal states.

8.73 a. p = the population proportion of correct answers.

H₀: $p = 0.50$ (he is just guessing), Hₐ: $p > 0.50$ (he is not just guessing).

One-proportion z-test.

b. p_a is the population proportion of athletes who can balance for at least 10 seconds. p_n is the population proportion of nonathletes who can balance for at least 10 seconds.

H₀: $p_a = p_n$, Hₐ: $p_a \neq p_n$

Two-proportion z-test.

8.75 H₀: $p = 0.50$, Hₐ: $p > 0.50$, $\hat{p} = \dfrac{13}{20} = 0.65$, $z = \dfrac{\hat{p} - p_0}{\sqrt{\dfrac{p_0(1 - p_0)}{n}}} = \dfrac{0.65 - 0.5}{\sqrt{\dfrac{0.5(0.5)}{20}}} = 1.34$, p-value = 0.09.

p : Proportion of successes
H₀ : p = 0.5
Hₐ : p > 0.5

Hypothesis test results:

Proportion	Count	Total	Sample Prop.	Std. Err.	Z-Stat	P-value
p	13	20	0.65	0.1118034	1.3416408	0.0899

Do not reject H₀. We have no evidence the student can tell tap water from bottled water.

8.77 0.05 (because $1 - 0.95 = 0.05$).

8.79 5% of 300, or 15.

8.81 Pew reports on all adults, the entire population, and so inference is not needed or appropriate. Also, rates have been given instead of counts.

8.83 a.   Step 1: H₀: $p_{2012} = p_{2018}$, Hₐ: $p_{2012} \neq p_{2018}$.

Step 2: Two-proportion z-test, $\hat{p} = \dfrac{66 + 76}{100 + 100} = \dfrac{142}{200}$, $= 0.71$; $n_1 \hat{p}_1 = 100(0.71)$, $n_1(1 - \hat{p}_1) = 100(0.29)$, $n_2 \hat{p}_2 = 100(0.71)$, $n_2(1 - \hat{p}_2) = 100(0.29)$; all expected counts are greater than 10.

Samples are random and assumed independent.

Step 3: Significance level = 0.01; $z = -1.56$, p-value = 0.12.

$$z = \frac{(\hat{p}_1 - \hat{p}_2 - 0)}{\sqrt{\hat{p}(1 - \hat{p})\left(\dfrac{1}{n_1} + \dfrac{1}{n_2}\right)}} = \frac{\left(\dfrac{66}{100} - \dfrac{76}{100} - 0\right)}{\sqrt{\dfrac{142}{200}\left(1 - \dfrac{142}{200}\right)\left(\dfrac{1}{100} + \dfrac{1}{100}\right)}} = -1.56$$

p_1 : proportion of successes for population 1
p_2 : proportion of successes for population 2
$p_1 - p_2$: Difference in proportions
$H_0 : p_1 - p_2 = 0$
$H_A : p_1 - p_2 \neq 0$

Hypothesis test results:

Difference	Count1	Total1	Count2	Total2	Sample Diff.	Std. Err.	Z-Stat	P-value
$p_1 - p_2$	66	100	76	100	-0.1	0.064171645	-1.5583207	0.1192

Step 4: Do not reject H_0. We cannot conclude the proportion of people who reported using Facebook was different in 2012 and 2018.

b.   Step 1: H_0: $p_{2012} = p_{2018}$, H_a: $p_{2012} \neq p_{2018}$.

Step 2: Two-proportion z-test, $\hat{p} = \dfrac{990 + 1140}{1500 + 1500} =, 0.71$; $n_1 \hat{p}_1 = 1500(0.71)$, $n_1(1 - \hat{p}_1) = 1500(0.29)$,

$n_2 \hat{p}_2 = 1500(0.71)$, $n_2(1 - \hat{p}_2) = 1500(0.29)$; all expected counts are greater than 10. Samples are random and assumed independent.

Step 3: Significance level = 0.01; $z = $ -6.04, p-value is approximately 0.

$$z = \frac{\left(\hat{p}_1 - \hat{p}_2 - 0 \right)}{\sqrt{\hat{p}\left(1 - \hat{p}\right)\left(\dfrac{1}{n_1} + \dfrac{1}{n_2}\right)}} = \frac{\left(\dfrac{990}{1500} - \dfrac{1140}{1500} - 0\right)}{\sqrt{\dfrac{2130}{3000}\left(1 - \dfrac{2130}{3000}\right)\left(\dfrac{1}{1500} + \dfrac{1}{1500}\right)}} = -6.04$$

p_1 : proportion of successes for population 1
p_2 : proportion of successes for population 2
$p_1 - p_2$: Difference in proportions
$H_0 : p_1 - p_2 = 0$
$H_A : p_1 - p_2 \neq 0$

Hypothesis test results:

Difference	Count1	Total1	Count2	Total2	Sample Diff.	Std. Err.	Z-Stat	P-value
$p_1 - p_2$	990	1500	1140	1500	-0.1	0.016569047	-6.03535	<0.0001

Step 4: Reject H_0. The proportions of Facebook users were different in 2012 and 2018.

c.   With the larger sample size (more evidence) we got a smaller p-value and were able to reject H_0.

8.85   It would not be appropriate to do such a test because the data were for the entire population of people who voted. Since the data are for a population, no inference is needed.

8.87   a.   Step 1: H_0: $p = 0.101$, H_a: $p \neq 101$.

Step 2: One- proportion z-test; $np = 500(0.101)$, $n(1 - p) = 500(.899)$, all expected counts greater than 10; sample random and assumed independent; large population: # workers assumed > 10(500).

Step 3: Significance level: 0.05; $z = 1.71$, p-value = 0.09.

$$\hat{p} = \frac{62}{500} = 0.124, \quad z = \frac{\hat{p} - p_0}{\sqrt{\dfrac{p_0(1 - p_0)}{n}}} = \frac{0.124 - 0.101}{\sqrt{\dfrac{0.101(0.899)}{500}}} = 1.71$$

p : Proportion of successes
$H_0 : p = 0.101$
$H_A : p \neq 0.101$

Hypothesis test results:

Proportion	Count	Total	Sample Prop.	Std. Err.	Z-Stat	P-value
p	62	500	0.124	0.01347583	1.7067594	0.0879

Step 4: Do not reject H_0. We cannot conclude the proportion of self-employed workers in his area is different from 10.1%.

b. (0.095, 0.153). This interval supports the hypothesis test conclusion because it contains 0.101.

p : Proportion of successes
Method: Standard-Wald

95% confidence interval results:

Proportion	Count	Total	Sample Prop.	Std. Err.	L. Limit	U. Limit
p	62	500	0.124	0.014739335	0.095111434	0.15288857

8.89 a. Quinnipiac: $824/1249 = 0.66$; NPR: $754/1005 = 0.75$

b. Step 1: H_0: $p_{Gallup} = p_{NBC}$; H_a: $p_{Gallup} \neq p_{NBC}$.

Step 2: Two-proportion z-test. $\hat{p} = \dfrac{824 + 754}{1249 + 1005} =, 0.70$; $n_1 \hat{p}_1 = 1249(0.70)$, $n_1(1 - \hat{p}_1) = 1249(0.30)$, $n_2 \hat{p}_2 = 1005(0.70)$, $n_2(1 - \hat{p}_2) = 1005(0.70)$; all expected counts are greater than 10. Samples are random and assumed independent.

Step 3: Significance level: 0.05, $z = -4.66$, p-value approximately 0.

$$z = \frac{(\hat{p}_1 - \hat{p}_2 - 0)}{\sqrt{\hat{p}(1-\hat{p})\left(\dfrac{1}{n_1} + \dfrac{1}{n_2}\right)}} = \frac{\left(\dfrac{824}{1249} - \dfrac{754}{1005} - 0\right)}{\sqrt{\dfrac{1578}{2254}\left(1 - \dfrac{1578}{2254}\right)\left(\dfrac{1}{1249} + \dfrac{1}{1005}\right)}} = -4.66$$

p₁ : proportion of successes for population 1
p₂ : proportion of successes for population 2
p₁ - p₂ : Difference in proportions
H₀ : p₁ - p₂ = 0
Hₐ : p₁ - p₂ ≠ 0

Hypothesis test results:

Difference	Count1	Total1	Count2	Total2	Sample Diff.	Std. Err.	Z-Stat	P-value
p₁ - p₂	824	1249	754	1005	-0.090520974	0.019417157	-4.6619067	<0.0001

Step 4: Reject H_0. The population proportions are not equal.

c. (–0.128, –0.053). The interval does not contain 0, supporting a conclusion that the population proportions are not equal. The difference between the population proportions is between 5.3% and 12.8%.

p₁ : proportion of successes for population 1
p₂ : proportion of successes for population 2
p₁ - p₂ : Difference in proportions

95% confidence interval results:

Difference	Count1	Total1	Count2	Total2	Sample Diff.	Std. Err.	L. Limit	U. Limit
p₁ - p₂	824	1249	754	1005	-0.090520974	0.019135746	-0.12802635	-0.053015602

8.91 a. The misconduct rate was higher for those in the sample who did not have three strikes $\left(\hat{p}_{others} = 974/3188 = 30.6\%\right)$ than for those in the sample who had three strikes $\left(\hat{p}_{3\text{-strikes}} = 163/734 = 22.2\%\right)$. This was not what was expected.

b. Step 1: H_0: $p_{3\text{-strikes}} = p_{others}$, H_a: $p_{3\text{-strikes}} > p_{others}$.

Step 2: Two-proportion z-test, $\alpha = 0.05$, $\hat{p} = \dfrac{974 + 163}{3188 + 734} = \dfrac{1137}{3922}$,

$$n_{3-strikes}\,\hat{p} = 734\left(\frac{1137}{3922}\right) = 213 > 10, \quad n_{3-strikes}\left(1 - \hat{p}\right) = 734\left(\frac{2785}{3922}\right) = 521 > 10,$$

$$n_{others}\,\hat{p} = 3188\left(\frac{1137}{3922}\right) = 924 > 10, \quad n_{others}\left(1 - \hat{p}\right) = 3188\left(\frac{2785}{3922}\right) = 2264 > 10$$

Random Assignment, Independence, both within samples and between samples, and Large Population are assumed.

Step 3:

$$z = \frac{\left(\hat{p}_1 - \hat{p}_2 - 0\right)}{\sqrt{\hat{p}(1-\hat{p})\left(\dfrac{1}{n_1} + \dfrac{1}{n_2}\right)}} = \frac{\left(\dfrac{163}{734} - \dfrac{974}{3188} - 0\right)}{\sqrt{\dfrac{1137}{3922}\left(1 - \dfrac{1137}{3922}\right)\left(\dfrac{1}{734} + \dfrac{1}{3188}\right)}} - 4.49,$$

(4.49 if proportions are reversed), p-value > 0.999

Test and CI for Two Proportions
Difference = p (1) - p (2)
Estimate for difference: -0.0834499
95% upper bound for difference: -0.0548693
Test for difference = 0 (vs < 0): Z = -4.49   P-Value = 0.000

Step 4: Do not reject H_0. The three strikers do not have a greater rate of misconduct than the other prisoners. (If a two-sided test had been done, the p-value would have been < 0.001, and we would have rejected the null hypothesis because the three-strikers had less misconduct.)

8.93   Step 1: H_0: $p = 0.52$, H_a: $p < 0.52$.

Step 2: One-proportion z-test. $np = 1024(0.52)$, $n(1 - p) = 1024(0.48)$, all expected counts greater than 10; sample random and assumed independent; large population: #Americans > 10(1024).

Step 3: Significance level: 0.05, $z = -4.47$, p-value is approximately 0.

$$\hat{p} = \frac{461}{1024} = 0.45, \quad z = \frac{\hat{p} - p_0}{\sqrt{\dfrac{p_0(1-p_0)}{n}}} = \frac{0.45 - 0.52}{\sqrt{\dfrac{0.52(0.48)}{1024}}} = -4.47$$

p : Proportion of successes
H_0 : p = 0.52
H_A : p < 0.52

Hypothesis test results:

Proportion	Count	Total	Sample Prop.	Std. Err.	Z-Stat	P-value
p	461	1024	0.45019531	0.015612495	-4.4710783	<0.0001

Step 4: Reject H_0. Satisfaction with the quality of the environment among Americans has decreased.

8.95   a   i; the chance of randomly picking the correct outcome is $1/4$.

b.   iii; you friend claims that he or she can do better than randomly guessing, $P > 1/4$.

8.97 $\hat{p} = \dfrac{12}{20} = 0.60, \quad z = \dfrac{\hat{p} - p_0}{\sqrt{\dfrac{p_0(1-p_0)}{n}}} = \dfrac{0.60 - 0.50}{\sqrt{\dfrac{0.50(0.50)}{20}}} = 0.89$

8.99   The p-value tells us that if the true proportion of those who text while driving is 0.25, then there is only a 0.034 probability that one would get a sample proportion of 0.125 or smaller with a sample size of 40.

8.101 He has not demonstrated ESP; 10 right out of 20 is only 50% right, which you should expect from guessing.

8.103 H_0: The death rate after starting hand washing is still 9.9%, or $p = 0.099$ (p is the proportion of all deaths at the clinic.)

H_a: The death rate after starting hand washing is less than 9.9%, or $p < 0.099$.

8.105 Step 3: $\hat{p} = \dfrac{35}{50} = 0.70$, $z = \dfrac{\hat{p} - p_0}{\sqrt{\dfrac{p_0(1-p_0)}{n}}} = \dfrac{0.70 - 0.50}{\sqrt{\dfrac{0.50(0.50)}{50}}} = 2.83$, p-value $= 0.002$

Test and CI for One Proportion
```
Test of p = 0.5 vs p > 0.5
                         95% Lower
Sample   X   N  Sample p    Bound   Z-Value  P-Value
1       35  50  0.700000  0.593401    2.83    0.002
Using the normal approximation.
```

Step 4: Reject H_0. The probability of doing this well by chance alone is so small that we must conclude that the student is not guessing.

Chapter 9: Inferring Population Means

Answers may vary slightly due to type of technology or rounding.

Section 9.1: Sample Means of Random Samples

9.1 a.   They are parameters, because they are for all the students, not a sample.

   b. $\mu = 20.7, \ \sigma = 2.5$

9.3 a.

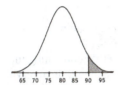

   b.   2.5% (By the Empirical Rule, 95% of the observed values are between 70 and 90 which leave 5% outside of these boundaries. We only want the right half, so the answer is half of 5%, 2.5%.)

9.5 a.   Answers will vary. We might expect it to be right-skewed if we think most people take relatively short showers and a few take very long showers. Some might expect it to be Normal because they think very few people take short showers and very few people take long showers. The summary statistics do not contradict a Normal distribution since it is possible to have 99% of the observations within three SDs of the mean. It is unlikely the distribution is left-skewed because the majority of people probably do not take extremely long showers.

   b.   By the Central Limit Theorem for sample means, the distribution of sample means will be approximately Normal because the sample size is greater than 25.

   c.   The mean will be the same as the population mean (8.2 minutes) and the standard deviation of the sample means will be $\sigma / \sqrt{n} = \ 2 / \sqrt{100} = 0.2$.

9.7 The distribution of *a* sample (*one* sample).

9.9 a.   22,306 miles

   b.   standard error $= \sigma / \sqrt{n} = 5500 / \sqrt{200} = 388.9$

Section 9.2: The Central Limit Theorem for Sample Means

9.11 a.   68% from the Empirical Rule. 6.4 is one standard deviation $\left(z = (6.4 - 7.0) / 0.6 = -1 \right)$ below the mean, and 7.6 is one standard deviation $\left(z = (7.6 - 7.0) / 0.6 = 1 \right)$ above the mean.

   b. $0.6 / \sqrt{4} = 0.3,$ so the standard error for the mean is 0.3. Now 7.6 is two standard errors $\left(z = (7.6 - 7.0) / 0.3 = -2 \right)$ above the mean, and 6.4 is two standard errors $\left(z = (6.4 - 7.0) / 0.3 = -2 \right)$ below the mean. The answer is 95% from the Empirical Rule.

   c.   The distribution of means is taller and narrower than the original distribution and will have more data in the central area.

9.13 a.   Yes; the sample is large, $n = 100 > 25,$ and so the distribution of means will be Normal.

   b.   The mean is 75,847, and the standard error is $\sigma / \sqrt{n} = \ 32000 / \sqrt{100} = 3200.$

   c.   32% from the Empirical Rule (if 68% of the data lie within one standard deviation (error) of the mean, then 32% lie beyond one standard deviation (error) of the mean.

9.15 Figure B is the original distribution; it is the least Normal and widest. Figure A is from samples of 5. Figure C is from samples of 25; it is the narrowest and the least skewed. The larger the sample size, the narrower and more Normal is the sampling distribution.

9.17 a. $\mu = 3.1$ (population mean), $\sigma = 2.7$ (population standard deviation), $\bar{x} = 2.7$ (sample mean) and
         $s = 2.1$ (sample standard deviation)

      b. $\mu = 3.1$ (population mean) and $\sigma = 2.7$ (population standard deviation) are parameters describing a
         population; $\bar{x} = 2.7$ (sample mean) and $s = 2.1$ (sample standard deviation) are statistics describing a
         sample.

      c.   Yes, sample is random and sample size (35) is greater than 25; the shape will be Normal.

Section 9.3: Answering Questions about the Mean of a Population

9.19 a.   (i) is incorrect because a confidence interval is for a population parameter, not a sample statistic;
         (ii) We are 95% confident that the population mean is between 3.71 and 3.79.
         (iii) is incorrect because the confidence level is not the probability of the interval containing the
         population parameter. It is a measure of our confidence in the method used to obtain the interval. Note: If
         using a t-distribution table, confidence interval limits are 3.61 and 3.89.

μ : Mean of population

95% confidence interval results:

Mean	Sample Mean	Std. Err.	DF	L. Limit	U. Limit
μ	3.75	0.018973666	9	3.7070786	3.7929214

      b.   The population mean GPA 3.80 falls just outside the upper limit so it is possible but unlikely. Note: if
         using a t-distribution table the population mean GPA 3.80 does fall in the interval and would.

9.21 *Random Sample, Independence, Big Population,* and *Large Sample* (*Normal Distribution*) are given.

      a.   i. is correct: (10.125, 10.525). Both ii. and iii. are incorrect.

$$x \pm t^* \frac{s}{\sqrt{n}} = 10.325 \pm 3.182 \left(\frac{0.1258}{\sqrt{4}} \right) = 10.325 \pm 0.200$$

```
Inverse Cumulative Distribution Function
Student's t distribution with 3 DF
P( X <= x )        x
   0.975   3.18245
One-Sample T: Orange Weight
Variable        N    Mean   StDev  SE Mean       95% CI
Orange Weight   4  10.3250  0.1258  0.0629  (10.1248, 10.5252)
```

      b.   No, it does not capture 10. Reject the claim of 10 pounds, because 10 is not in the interval.

9.23 a.   is incorrect because the confidence level is not the probability of the interval containing the population
         parameter. It is a measure of our confidence in the method used to obtain the interval.

      b.   is the correct interpretation. a. is incorrect because the confidence level is not the probability of the
         interval containing the population parameter. It is a measure of our confidence in the method used to
         obtain the interval.

9.25 Use $t^* = 2.056$.

9.27 a.   Sample size: 30; Sample mean: 170.7; Standard deviation: 11.5

      b.   I am 95% confident that the population mean height of 12th grade students is between 166 and 175cm.

9.29 *Random Sample, Independence,* and *Big Population* are met. *Large Sample* (*Normal Distribution*):
      $n = 25 \geq 25$.

      a.   (67, 77); I am 95% confident that the mean is between 67 and 77 beats per minute.

$$x \pm t^* \frac{s}{\sqrt{n}} = 72 \pm 2.064 \left(\frac{13}{\sqrt{25}} \right) = 72 \pm 5.37; \ \ 72 - 5.37 = 66.63; \ \ 72 + 5.37 = 77.37$$

```
One-Sample T
  N   Mean  StDev  SE Mean      95% CI
 25  72.00  13.00    2.60   (66.63, 77.37)
```

b. (65, 79); I am 99% confident that the mean is between 65 and 79 beats per minute.

$$x \pm t * \frac{s}{\sqrt{n}} = 72 \pm 2.797 \left(\frac{13}{\sqrt{25}} \right) = 72 \pm 7.27; \ 72 - 7.27 = 64.73; \ 72 + 7.27 = 79.27$$

```
One-Sample T
  N   Mean  StDev  SE Mean      99% CI
 25  72.00  13.00    2.60   (64.73, 79.27)
```

c. The 99% interval is wider because a greater level of confidence requires a bigger value for $t *$.

9.31 a. (22.0, 42.8) is the 90% interval because it is narrower than (19.9, 44.0).

b. If a larger sample size was used, the intervals would be smaller because the standard error of the distribution would be smaller.

9.33 a. Wider: because a greater level of confidence requires a larger $t *$.

b. Narrower: A larger sample size produces a narrower interval.

c. Wider: A larger standard deviation increases the standard error.

9.35 *Random Sample*, *Independence*, and *Big Population* are met. *Large Sample* (*Normal Distribution*) is given.

a. (20.4, 21.7); I am 95% confident that the population mean is between 20.4 and 21.7 pounds.

$$\bar{x} \pm t * \frac{s}{\sqrt{n}} = 21.05 \pm 3.182 \left(\frac{0.420}{\sqrt{4}} \right) = 21.05 \pm 0.668$$

```
Inverse Cumulative Distribution Function
Student's t distribution with 3 DF
P( X <= x )         x
     0.975    3.18245
One-Sample T: Potato Weight
Variable        N    Mean  StDev  SE Mean       95% CI
Potato Weight   4  21.050  0.420    0.210  (20.381, 21.719)
```

b. The interval does not capture 20 pounds. There is enough evidence to reject 20 pounds as the population mean.

Section 9.4: Hypothesis Testing for Means

9.37 The answers follow the guidance on page 474.

Step 1: H₀: $\mu = 98.6$, Hₐ: $\mu \neq 98.6$, where μ is the population mean body temperature.

Step 2: One-sample *t*-test: *Random Sample*, *Independence*, and *Big Population* are met. *Large Sample* (*Normal Distribution*): $n = 10 < 25$, but the data not very skewed (see stemplot below). $\alpha = 0.05$.

```
Stem-and-Leaf Display: Body Temp
Stem-and-leaf of Body Temp   N  = 10
Leaf Unit = 0.10
  1   96   3
  1   96
  3   97   22
  3   97
  5   98   23
  5   98   577
  2   99   01
```

Step 3: $t = -1.65$, p-value $= 0.133$.

$$t = \frac{\bar{x} - \mu}{s / \sqrt{n}} = \frac{98.120 - 98.6}{0.919 / \sqrt{10}} = -1.65$$

9.37 (cont.)

One-Sample T: Body Temp
```
Test of mu = 98.6 vs not = 98.6
Variable    N    Mean   StDev  SE Mean      95% CI          T      P
Body Temp  10  98.120  0.919   0.291  (97.463, 98.777)  -1.65  0.133
```

Step 4: Do not reject H_0. Choose i.

9.39 a.   You should be able to reject 20 pounds because the confidence interval (20.4 to 21.7) did not capture 20 pounds.

b.   Step 1: H_0: $\mu = 20$ H_a: $\mu \neq 20$.

Step 2: One-sample *t*-test: *Random Sample*, *Independence*, and *Big Population* are met. *Large Sample* (*Normal Distribution*) is given. $\alpha = 0.05$

Step 3: $t = 5.00$, p-value $= 0.015$.

$$t = \frac{\bar{x} - \mu}{s / \sqrt{n}} = \frac{21.05 - 20}{0.420 / \sqrt{4}} = 5.00$$

One-Sample T: Potato Weight
```
Test of mu = 20 vs not = 20
Variable        N    Mean   StDev  SE Mean      95% CI          T      P
Potato Weight   4  21.050  0.420   0.210  (20.381, 21.719)  5.00  0.015
```

c.   Yes, reject H_0. Choose ii.

9.41 Step 1: H_0: $\mu = 200$, H_a: $\mu > 200$, where μ is the population mean cholesterol level.

Step 2: One-sample *t*-test: *Random Sample*, *Independence*, and *Big Population* are met. *Large Sample* (*Normal Distribution*): $n = 50 > 25$, $\alpha = 0.05$.

Step 3: $t = 1.44$, p-value $= 0.079$.

$$t = \frac{\bar{x} - \mu}{s / \sqrt{n}} = \frac{208.26 - 200}{40.67 / \sqrt{50}} = 1.44$$

Step 4: Do not reject H_0. We have not shown that the mean is significantly more than 200.

9.43 a.   Step 1: H_0: $\mu = 38$, H_a: $\mu \neq 38$, where μ is the population mean height of 3-year-old boys.

Step 2: One-sample *t*-test: *Random Sample*, *Independence*, *Big Population*, and *Large Sample* (*Normal Distribution*) are given, $\alpha = 0.05$.

Step 3: $t = -1.03$, p-value $= 0.319$.

$$t = \frac{\bar{x} - \mu}{s / \sqrt{n}} = \frac{37.2 - 38}{3 / \sqrt{15}} = -1.03$$

One-Sample T
```
Test of mu = 38 vs not = 38
 N    Mean   StDev  SE Mean      95% CI          T      P
15  37.200  3.000   0.775  (35.539, 38.861)  -1.03  0.319
```

Step 4: Do not reject H_0. The mean for non-U.S. boys is not significantly different from 38.

b.   Step 3: $t = -1.46$, p-value $= 0.155$.

$$t = \frac{\bar{x} - \mu}{s / \sqrt{n}} = \frac{37.2 - 38}{3 / \sqrt{30}} = -1.46$$

```
One-Sample T
Test of mu = 38 vs not = 38
 N    Mean   StDev   SE Mean      95% CI        T      P
 30   37.200  3.000   0.548  (36.080, 38.320)  -1.46  0.155
```

Step 4: Do not reject H$_0$. The mean for non-U.S. boys is not significantly different from 38.

c. Larger *n*, smaller standard error (narrower sampling distribution) with less area in the tails, as shown by the smaller p-value.

9.45 a. *Step 1*: H$_0$: $\mu = 12.5$, H$_a$: $\mu < 12.5$, where μ is the population mean psi.

Step 2: One sample *t*-test: Assume *Random Sample*, *Independence*, and *Normal Distribution*:

Step 3: $\alpha = 0.05$. $t = -11.46$, p-value <0.001.

μ : Mean of variable
H$_0$: μ = 12.5
H$_A$: μ < 12.5

Hypothesis test results:

Variable	Sample Mean	Std. Err.	DF	T-Stat	P-value
psi	11.109091	0.12132192	10	-11.464615	<0.0001

Step 4: Reject H$_0$. The pressure is less than 12.5 psi. This shows the balls are deflated.

b. (10.84.11.38); the interval does not contain 12.5, only values < 12.

μ : Mean of variable

95% confidence interval results:

Variable	Sample Mean	Std. Err.	DF	L. Limit	U. Limit
psi	11.109091	0.12132192	10	10.838769	11.379413

9.47 a. Step 1: H$_0$: $\mu = 8.97$, H$_a$: $\mu \neq 8.97$ where μ is the population mean for a movie ticket in 2018.

Step 2: One sample *t*-test: $n \geq 25$, Assume *Random Sample*, *Independence*, and *Normal Distribution*:

Step 3: $\alpha = 0.05$. $t = 4.91$, p-value = approximately 0.

μ : Mean of population
H$_0$: μ = 8.97
H$_A$: μ ≠ 8.97

Hypothesis test results:

Mean	Sample Mean	Std. Err.	DF	T-Stat	P-value
μ	12.27	0.672	24	4.9107143	<0.0001

Step 4: Reject H$_0$, the mean price of a movie ticket in the San Francisco Bay Area is different from the national average.

b. (10.88, 13.66); The interval does not include 8.97, confirming the mean price is different from the national average.

μ : Mean of population

95% confidence interval results:

Mean	Sample Mean	Std. Err.	DF	L. Limit	U. Limit
μ	12.27	0.672	24	10.88306	13.65694

9.49 Step 1: H$_0$: $\mu = 0$, H$_a$: $\mu > 0$, where μ is the population mean weight loss.

Step 2: One-sample *t*-test: *Random Sample* not met (don't generalize), *Independence*, and *Big Population* are met. *Large Sample* (*Normal Distribution*): given, $\alpha = 0.05$.

Step 3: $t = 3.60$, p-value = 0.003.

$$t = \frac{\overline{x} - \mu}{s/\sqrt{n}} = \frac{4.1 - 0}{3.604/\sqrt{10}} = 3.60$$

One-Sample T: Weight Loss
Test of mu = 0 vs > 0

Variable	N	Mean	StDev	SE Mean	95% Lower Bound	T	P
Weight Loss	10	4.10	3.60	1.14	2.01	3.60	0.003

Step 4: Reject H_0. There was a significant weight loss, but don't generalize.

9.51 You would expect $0.95(200) = 190$ to capture and $200 - 190 = 10$ to miss.

Section 9.5: Comparing Two Population Means

9.53 a. Paired

b. Independent

9.55 a. The samples are random, independent, and large $(n_1 = n_2 = 30 > 25)$, so the conditions are met.

b. I am 95% confident that the mean difference (OC – MC) is between –0.40 and 1.14 TVs.

c. The interval for the difference captures 0, which implies that it is plausible that the means are the same.

9.57 The answers follow the guidance on page 474.

Step 1: H_0: $\mu_{oc} = \mu_{mmc}$, H_a: $\mu_{oc} \neq \mu_{mmc}$, where μ is the population mean number of TVs.

Step 2: Two-sample t-test: *Random Samples and Independent Observations* and *Independent Samples* are given. *Large Sample (Normal Distribution)*: $n_1 = n_2 = 30 > 25$, $\alpha = 0.05$.

Step 3: $t = 0.95$, p-value = 0.345.

$$t = \frac{\bar{x}_1 - \bar{x}_2 - 0}{\sqrt{s_1^2 / n_1 + s_2^2 / n_2}} = \frac{0.367 - 0}{\sqrt{(1.49)^2 / 30 + (1.49)^2 / 30}} = 0.954 \text{ (or } -0.954 \text{ if samples reversed.)}$$

Step 4: Do not reject H_0. Choose i. Confidence interval: (–0.404, 1.138). Because the interval for the difference captures 0, we cannot reject the hypothesis that the mean difference in number of TVs is 0.

9.59 a. The men's sample mean triglyceride level of 139.5 was higher than the women's sample mean of 84.4.

b. Step 1: H_0: $\mu_{men} = \mu_{women}$, H_a: $\mu_{men} > \mu_{women}$, where μ is the population mean triglyceride level.

Step 2: Two-sample t-test: Assume that *Random Samples and Independent Observations* and *Independent Samples* are met. *Large Sample (Normal Distribution)*: $n_1 = 44 > 25, n_2 = 48 > 25$, $\alpha = 0.05$.

Step 3: $t = 4.02$ or -4.02, p-value < 0.001.

$$t = \frac{\bar{x}_1 - \bar{x}_2 - 0}{\sqrt{s_1^2 / n_1 + s_2^2 / n_2}} = \frac{84.4 - 139.5 - 0}{\sqrt{(40.2)^2 / 44 + (85.3)^2 / 48}} = 4.02 \text{ (or } -4.02 \text{ if samples reversed.)}$$

Step 4: Reject H_0. The mean triglyceride level is significantly higher for men than for women. Choose output B: Difference = $\mu_{men} - \mu_{women}$ which tests whether this difference is less than 0, and that is the one-sided hypothesis that we want.

9.61 $(-82.5, -27.7)$; because the difference of 0 is not captured, it shows there is a significant difference. Also, the difference $\mu_{female} - \mu_{male}$ is negative, which shows that the men's mean (triglyceride level) is significantly higher than the women's mean.

9.63 Step 1: H_0: $\mu_{American} = \mu_{National}$, H_a: $\mu_{American} \cdot \mu_{National}$.

Step 2: Two-sample t-test; random, independent, Normal;

Step 3: Significance level: 0.05, $t = -0.25$, p-value = 0.803.

Two sample T hypothesis test:

μ_1 : Mean of American
μ_2 : Mean of National
$\mu_1 - \mu_2$: Difference between two means
$H_0 : \mu_1 - \mu_2 = 0$
$H_A : \mu_1 - \mu_2 \neq 0$
(without pooled variances)

Hypothesis test results:

Difference	Sample Diff.	Std. Err.	DF	T-Stat	P-value
$\mu_1 - \mu_2$	-156.25747	616.99483	26.476249	-0.25325572	0.802

Step 4: Do not reject H_0. We cannot conclude there is a difference in mean salaries for the two leagues.

9.65   a.   Yes, the 95% CI would contain 0 since we cannot conclude there is a difference in mean salaries for the two leagues.

   b.   (−1423.40, 1110.88); the confidence interval does contain 0 which supports the hypothesis test conclusion.

Two sample T confidence interval:

μ_1 : Mean of American
μ_2 : Mean of National
$\mu_1 - \mu_2$: Difference between two means
(without pooled variances)

95% confidence interval results:

Difference	Sample Diff.	Std. Err.	DF	L. Limit	U. Limit
$\mu_1 - \mu_2$	-156.25747	616.99483	26.476249	-1423.3991	1110.8842

9.67   a. $\bar{x}_{UCSB} = \$61.01$ and $\bar{x}_{CSUN} = \$75.55$, so the sample mean at CSUN was larger.

   b.   Step 1: H_0: $\mu_{UCSB} = \mu_{CSUN}$, H_a: $\mu_{UCSB} \neq \mu_{CSUN}$, where μ is the population mean book price.

   Step 2: Paired t-test (matched pairs), *Random Samples and Independent Observations, Independent Samples*, and *Large Sample (Normal Distribution)* are given. $\alpha = 0.05$

   Step 3: $t = -3.21$ or 3.21, p-value = 0.004.

$$t_{Difference} = \frac{\bar{x}_{Difference} - 0}{s_{Difference}/\sqrt{n}} = \frac{-14.54 - 0}{22.20/\sqrt{24}} = -3.21 \text{ (or 3.21 if difference is reversed.)}$$

```
Paired T-Test and CI: UCSB, CSUN
Paired T for UCSB - CSUN
                N    Mean  StDev  SE Mean
UCSB           24   61.01  31.66     6.46
CSUN           24   75.55  33.89     6.92
Difference     24  -14.54  22.20     4.53
95% CI for mean difference: (-23.92, -5.17)
T-Test of mean difference = 0 (vs not = 0): T-Value = -3.21  P-Value = 0.004
```

Step 4: You can reject H_0. The means are significantly different.

9.69   The answers follow the guidance on page 475.

Step 1: H_0: $\mu_{before} = \mu_{after}$, H_a: $\mu_{before} < \mu_{after}$, where μ is the population mean pulse rate.

Step 2: Paired t-test: each woman is measured twice (repeated measures), so a measurement in the first column is coupled with a measurement of the same person in the second column. *Random Sample, Independence*, and *Large Sample (Normal Distribution)* are given. $\alpha = 0.05$

Step 3: $t = 4.90$ or –4.90, p-value < 0.001.

$$t_{Difference} = \frac{\bar{x}_{Difference} - 0}{s_{Difference}/\sqrt{n}} = \frac{-8.92 - 0}{6.56/\sqrt{13}} = -4.90 \text{ (or 4.90 if difference is reversed.)}$$

```
Paired T-Test and CI: Fem_Before, Fem_After
Paired T for Fem_Before - Fem_After
               N    Mean   StDev   SE Mean
Fem_Before    13   74.77   12.37     3.43
Fem_After     13   83.69   13.61     3.77
Difference    13   -8.92    6.56     1.82
95% upper bound for mean difference: -5.68
T-Test of mean difference = 0 (vs < 0): T-Value = -4.90   P-Value = 0.000
```

Step 4: Reject H_0. The sample mean before was 74.8, and the sample mean after was 83.7. The pulse rates of women go up significantly after they hear a scream.

9.71   Choose Figure B. The items are paired because the same items were priced at each store.

Step 1: H_0: $\mu_{target} = \mu_{wholefoods}$, H_a: $\mu_{target} \neq \mu_{wholefoods}$, where μ is the population mean price.

Step 2: Paired *t*-test: *Random Sample* and *Independence* are assumed. *Large Sample* (*Normal Distribution*): $n = 30 > 25$, $\alpha = 0.05$.

Step 3: $t = -1.26$ or 1.26, p-value = 0.217.

$$t_{Difference} = \frac{\bar{x}_{Difference} - 0}{s_{Difference} / \sqrt{n}} = \frac{-0.265 - 0}{1.152 / \sqrt{30}} = -1.26 \text{ (or 1.26 if difference is reversed.)}$$

Step 4: Do not reject H_0. The means are not significantly different.

9.73   a.   Step 1: H_0: $\mu_{ales} = \mu_{IPAs}$, H_a: $\mu_{ales} \neq \mu_{IPAs}$

Step 2: 2-sample *t*-test; assume random, Normal;

Step 3: Significance level: 0.05, $t = -2.57$, p-value = 0.025.

μ_1 : Mean of Population 1
μ_2 : Mean of Population 2
$\mu_1 - \mu_2$: Difference between two means
$H_0 : \mu_1 - \mu_2 = 0$
$H_A : \mu_1 - \mu_2 \neq 0$
(without pooled variances)

Hypothesis test results:

Difference	Sample Diff.	Std. Err.	DF	T-Stat	P-value
$\mu_1 - \mu_2$	-57.5	22.23445	11.662527	-2.586077	0.0243

Step 4: Reject H_0. There is a difference between the mean calorie content of ales and IPAs.

b.   Step 1: H_0: $\mu_{ales} = \mu_{IPAs}$, H_a: $\mu_{ales} < \mu_{IPAs}$.

Step 2: 2-sample *t*-test; assume random, Normal;

Step 3: Significance level: 0.05; $t = -2.57$, p-value = 0.013.

μ_1 : Mean of Population 1
μ_2 : Mean of Population 2
$\mu_1 - \mu_2$: Difference between two means
$H_0 : \mu_1 - \mu_2 = 0$
$H_A : \mu_1 - \mu_2 < 0$
(without pooled variances)

Hypothesis test results:

Difference	Sample Diff.	Std. Err.	DF	T-Stat	P-value
$\mu_1 - \mu_2$	-57.1	22.23445	11.662527	-2.5680869	0.0126

Step 4: Reject H_0. The mean caloric content of IPAs is significantly larger than that of Ales. The p-value is about half the p-value of the two-tailed test.

9.75   a.   The 95% confidence interval, $(-1.44, 0.25)$, captures 0, so the hypothesis that the means are equal cannot be rejected.

$$\bar{x}_{Difference} \pm t * \frac{s_{Difference}}{\sqrt{n}} = -0.595 \pm 2.145 \left(\frac{1.533}{\sqrt{15}} \right) = -0.595 \pm 0.849$$

b.   Step 1: H₀: $\mu_{\text{measured}} = \mu_{\text{reported}}$, Hₐ: $\mu_{\text{measured}} \neq \mu_{\text{reported}}$, where μ is population mean height of men.

Step 2: Paired t-test: each person is the source of two numbers. *Random Samples and Independent Observations* and *Independent Samples* are given. $\alpha = 0.05$

Step 3: $t = 1.50$ or -1.50, p-value $= 0.155$.

$$t_{\text{Difference}} = \frac{\overline{x}_{\text{Difference}} - 0}{s_{\text{Difference}} / \sqrt{n}} = \frac{-0.595 - 0}{1.533 / \sqrt{15}} = -1.50 \text{ (or 1.50 if difference is reversed.)}$$

```
Inverse Cumulative Distribution Function
Student's t distribution with 14 DF
P( X <= x )        x
   0.975    2.14479
Paired T-Test and CI: Reported, Measured
Paired T for Reported - Measured
                N     Mean  StDev  SE Mean
Reported       15   176.87   7.96     2.06
Measured       15   177.46   8.02     2.07
Difference     15   -0.595  1.533    0.396
95% CI for mean difference: (-1.444, 0.254)
T-Test of mean difference = 0 (vs not = 0): T-Value = -1.50   P-Value = 0.155
```

Step 4: Do not reject H₀. The mean measured and reported heights are not significantly different for men or there is not enough evidence to support the claim that the typical self-reported height differs from the typical measured height for men.

9.77 a.   (−1.67, 3.24). Since the interval contains 0, there is not a significant difference in the heart rates of females and males.

Two sample T confidence interval:

μ_1 : Mean of Female Heart Rate
μ_2 : Mean of Male Heart Rate
$\mu_1 - \mu_2$: Difference between two means
(without pooled variances)

95% confidence interval results:

Difference	Sample Diff.	Std. Err.	DF	L. Limit	U. Limit
$\mu_1 - \mu_2$	0.78461538	1.2416645	116.70438	-1.6745012	3.243732

b.   Step 1: H₀: $\mu_{\text{males}} = \mu_{\text{females}}$, Hₐ: $\mu_{\text{males}} \neq \mu_{\text{females}}$;

Step 2: 2-sample t-test; assume random, Normal;

Step 3: Significance level: 0.05; $t = 0.63$, p-value $= 0.53$.

μ_1 : Mean of Female Heart Rate
μ_2 : Mean of Male Heart Rate
$\mu_1 - \mu_2$: Difference between two means
H₀ : $\mu_1 - \mu_2 = 0$
Hₐ : $\mu_1 - \mu_2 \neq 0$
(without pooled variances)

Hypothesis test results:

Difference	Sample Diff.	Std. Err.	DF	T-Stat	P-value
$\mu_1 - \mu_2$	0.78461538	1.2416645	116.70438	0.6319061	0.5287

Step 4: Do not Reject H₀. We cannot conclude there is a significant difference in the mean heart rate of females and males.

Chapter Review Exercises

9.79 a.   The distribution is normal so the question can be answered. N(65, 2.5), $p(x < 63) = 0.212$

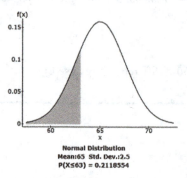

Normal Distribution
Mean:65 Std. Dev.:2.5
P(X≤63) = 0.2118554

  b.   Since the population is normally distributed, we can apply the Central Limit Theorem. For samples of size 5, standard error $= \left(\dfrac{2.5}{\sqrt{5}}\right) = 1.12$; N(65, 1.12), $p(x < 63) = 0.037$

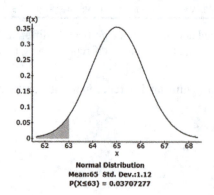

Normal Distribution
Mean:65 Std. Dev.:1.12
P(X≤63) = 0.03707277

  c.   Since the population is normally distributed, we can apply the Central Limit Theorem. For samples of size 5, standard error $= \left(\dfrac{2.5}{\sqrt{30}}\right) = 0.456$; N(65, 0.456), $p(x < 63) =$ approximately 0.

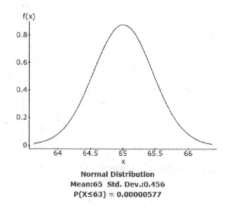

Normal Distribution
Mean:65 Std. Dev.:0.456
P(X≤63) = 0.00000577

9.81 a.   One sample *t*-test

  b.   Two-sample *t*-test

  c.   No *t*-test (two categorical variables)

9.83   *Random Sample*, *Independence*, and *Big Population* met. *Large Sample* (*Normal Distribution*) is given.

a. H_0: $\mu = 3.18$, H_a: $\mu \neq 3.18$, where μ is the population mean weight of ice cream cones. $t = 3.66$, p-value = 0.035. Reject H_0. The mean is significantly different from 3.18.

$$t = \frac{\bar{x} - \mu}{s/\sqrt{n}} = \frac{3.975 - 3.18}{0.43493/\sqrt{4}} = 3.66$$

One-Sample T: Cone Weight

Test of mu = 3.18 vs not = 3.18

Variable	N	Mean	StDev	SE Mean	95% CI	T	P
Cone Weight	4	3.975	0.435	0.217	(3.283, 4.667)	3.66	0.035

b. H_0: $\mu = 3.18$, H_a: $\mu < 3.18$, where μ is the population mean weight of ice cream cones. $t = 3.66$, p-value = 0.982, do not reject H_0. The mean is not significantly less than 3.18 ounces.

One-Sample T: Cone Weight

Test of mu = 3.18 vs < 3.18

Variable	N	Mean	StDev	SE Mean	95% Upper Bound	T	P
Cone Weight	4	3.975	0.435	0.217	4.487	3.66	0.982

c. H_0: $\mu = 3.18$, H_a: $\mu > 3.18$, where μ is the population mean weight of ice cream cones. $t = 3.66$, p-value = 0.018, reject H_0. The mean is significantly more than 3.18 ounces.

One-Sample T: Cone Weight

Test of mu = 3.18 vs > 3.18

Variable	N	Mean	StDev	SE Mean	95% Lower Bound	T	P
Cone Weight	4	3.975	0.435	0.217	3.463	3.66	0.018

9.85   Step 1: H_0: $\mu_{men} = \mu_{women}$, H_a: $\mu_{men} > \mu_{women}$, where μ is the population mean brain size.

Step 2: Two-sample *t*-test: *Random Samples and Independent Observations* and *Independent Samples* are assumed. *Large Sample* (*Normal Distribution*): The graphs do not appear too skewed, so it can be assumed that $n_1 = 20$ and $n_2 = 20$ are large enough. $\alpha = 0.05$

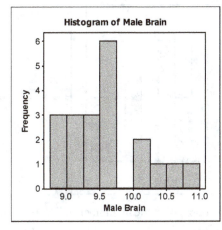
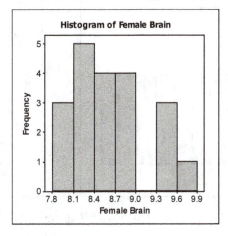

Step 3: $t = 5.27$ (or -5.27), p-value < 0.001.

$$t = \frac{\bar{x}_1 - \bar{x}_2 - 0}{\sqrt{s_1^2/n_1 + s_2^2/n_2}} = \frac{9.560 - 8.625 - 0}{\sqrt{(0.560)^2/20 + (0.561)^2/20}} = 5.28 \text{ (or } -5.28 \text{ if samples are reversed.)}$$

Two-Sample T-Test and CI: Male Brain, Female Brain
```
Two-sample T for Male Brain vs Female Brain
               N    Mean   StDev  SE Mean
Male Brain     20   9.560  0.560     0.13
Female Brain   20   8.625  0.561     0.13
Difference = mu (Male Brain) - mu (Female Brain)
Estimate for difference:  0.935
95% lower bound for difference:  0.636
T-Test of difference = 0 (vs >): T-Value = 5.27  P-Value = 0.000  DF = 37
```

Step 4: Reject H_0. The mean for brain size for men is significantly more than the mean for women.

9.87   Step 1: H_0: $\mu_{\text{before}} = \mu_{\text{after}}$, H_a: $\mu_{\text{before}} < \mu_{\text{after}}$, where μ is the population mean pulse rate.

Step 2: Paired *t*-test (repeated measures): *Random Sample, Independence, Big Population*, and *Large Sample* (*Normal Distribution*) are given.

Step 3: $t = 2.96$ or -2.96, p-value = 0.005.

$$t_{\text{Difference}} = \frac{\overline{x}_{\text{Difference}} - 0}{s_{\text{Difference}}/\sqrt{n}} = \frac{-5.07 - 0}{6.63/\sqrt{15}} = -2.96 \text{ (or 2.96 if difference is reversed.)}$$

Paired T-Test and CI: Before, After
```
Paired T for Before - After
             N    Mean   StDev  SE Mean
Before       15   82.40   9.33    2.41
After        15   87.47  10.18    2.63
Difference   15   -5.07   6.63    1.71
T-Test of mean difference = 0 (vs < 0): T-Value = -2.96  P-Value = 0.005
```

Step 4: Reject H_0. Heart rates increase significantly after coffee. (The average rate before coffee was 82.4, and the average rate after coffee was 87.5.)

9.89   The typical number of hours was a little higher for the boys, and the variation was almost the same. $\overline{x}_{\text{girls}} = 9.8$, $\overline{x}_{\text{boys}} = 10.3$, $s_{\text{girls}} = 5.4$, and $s_{\text{boys}} = 5.5$.

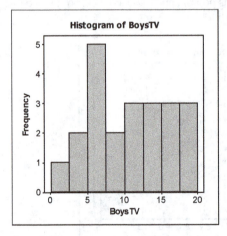
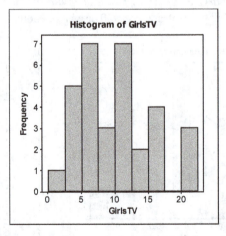

Step 1: H_0: $\mu_{\text{boys}} = \mu_{\text{girls}}$, H_a: $\mu_{\text{boys}} \neq \mu_{\text{girls}}$, where μ is the population mean number of TV viewing hours.

Step 2: Two-sample *t*-test: *Random Samples and Independent Observations* and *Independent Samples* are given. We must assume the sample sizes of 32 girls and 22 boys are large enough that the slight non-Normality seen in the histograms above is not a problem. $\alpha = 0.05$

Step 3: $t = -0.38$ or 0.38, p-value = 0.706.

$$t = \frac{\overline{x}_1 - \overline{x}_2 - 0}{\sqrt{s_1^2/n_1 + s_2^2/n_2}} = \frac{10.34 - 9.77 - 0}{\sqrt{(5.52)^2/22 + (5.38)^2/32}} = 0.38 \text{ (or } -0.38 \text{ if samples are reversed.)}$$

Two-Sample T-Test and CI: GirlsTV, BoysTV
```
Two-sample T for GirlsTV vs BoysTV
          N   Mean  StDev  SE Mean
GirlsTV   32   9.77   5.38    0.95
BoysTV    22  10.34   5.52    1.2
Difference = mu (GirlsTV) - mu (BoysTV)
Estimate for difference:  -0.58
95% upper bound for difference:  1.97
T-Test of difference = 0 (vs <): T-Value = -0.38  P-Value = 0.353  DF = 44
```

Step 4: You cannot reject the null hypothesis. There is not enough evidence to conclude that boys and girls differ in the typical hours of TV watched.

9.91 a. A two-sample t-test since the data are not paired.

b. 95% confidence interval for $\mu_{ales} - \mu_{lagers}$: (11.88, 47.76). Since the interval does not contain 0, there is a significant different in the mean number of calories contained in ales and lagers. Since both numbers in the confidence intervals are positive, we can conclude that ales contain more calories than lagers.

Two sample T summary confidence interval:

μ_1 : Mean of Population 1
μ_2 : Mean of Population 2
$\mu_1 - \mu_2$: Difference between two means
(without pooled variances)

95% confidence interval results:

Difference	Sample Diff.	Std. Err.	DF	L. Limit	U. Limit
$\mu_1 - \mu_2$	29.82	8.7599945	28.027987	11.876772	47.763228

9.93 a. Paired t-test because the data are paired (2 prices associated with one food item).

b. Step 1: H$_0$: $\mu_{difference} = 0$, where $\mu_{difference} = \mu_{Amazon} - \mu_{Walmart}$, H$_a$: $\mu_{difference} \neq 0$.

Step 2: Paired t-test, random, Normal populations;

Step 3: Significance level: 0.05, $t = -1.50$, p-value = 0.1622

$\mu_D = \mu_1 - \mu_2$: Mean of the difference between Amazon and Walmart
H$_0$: $\mu_D = 0$
H$_A$: $\mu_D \neq 0$
Hypothesis test results:

Difference	Mean	Std. Err.	DF	T-Stat	P-value
Amazon - Walmart	-0.14083333	0.093997569	11	-1.4982657	0.1622

Step 4: Do not reject H$_0$. We cannot conclude there is a difference in the mean prices for the two delivery services.

9.95 The table below shows the results. The average of s^2 in the table is $26/9 = 2.8889$ (or about 2.89), and if you take the square root, you get about 1.6997 (or about 1.70), which is the value for sigma (s) given in the TI-84 output shown in the exercise. This demonstrates that s^2 is an unbiased estimator of σ^2, sigma squared.

Sample	s	s^2
1, 1	0	0
1, 2	0.7071	0.5
1, 5	2.8284	8.0
2, 1	0.7071	0.5
2, 2	0	0
2, 5	2.1213	4.5
5, 1	2.8284	8.0
5, 2	2.1213	4.5
5, 5	0	0
Sum		26.0

9.97 Answers will vary.

Chapter 10: Associations between Categorical Variables

Answers may vary slightly due to type of technology or rounding.

Section 10.1: The Basic Ingredients for Testing with Categorical Variables

10.1 a. Proportions are used for categorical data.

b. Chi-square tests are used for categorical data.

10.3

	Boys	Girls
Violent	10	11
Nonviolent	19	4

The table may have a different orientation.

10.5 *Mean GPA*: numerical and continuous. *Field of Study*: categorical.

10.7 a. 72

b. $72 / 120 = 0.6$

c. $0.53(120) = 63.6$

10.9 a.

Eat Breakfast at least 3x weekly	Females	Males	Total
Yes	206	94	300
No	92	49	141
Total	298	143	441

b. $300 / 441 = 0.68 = 68\%$

c. Females: $0.68(298) = 202.64$, Males: $0.68(143) = 97.24$

d. Females: $0.32(298) = 95.36$, Males: $0.32(143) = 45.76$; expected counts shown in the table:

Eat Breakfast at least 3x weekly	Females	Males	Total
Yes	202.64	97.24	299.88
No	95.36	45.76	141.12
Total	298	143	441

e. $X^2 = \dfrac{(206 - 202.64)^2}{202.64} + \dfrac{(94 - 97.24)^2}{97.24} + \dfrac{(92 - 95.36)^2}{95.36} + \dfrac{(49 - 45.76)^2}{45.76} = 0.51$

10.11 a. $0.40(16) = 6.4$ should have had heart disease and $0.60(16) = 9.6$ should not.

b. $X^2 = \dfrac{(9 - 6.4)^2}{6.4} + \dfrac{(7 - 9.6)^2}{9.6} = \dfrac{6.76}{6.4} + \dfrac{6.76}{9.6} = 1.056 + 0.704 = 1.76$

Chi-Square Goodness-of-Fit Test for Observed Counts in Variable: Mummies

Category	Observed	Test Proportion	Expected	Contribution to Chi-Sq
1	9	0.4	6.4	1.05625
2	7	0.6	9.6	0.70417

N	DF	Chi-Sq	P-Value
16	1	1.76042	0.185

10.13 a. $0.2(152) = 30.4$ (expected correct), $0.8(152) = 121.6$ (expected incorrect).

b. $X^2 = \dfrac{(39-30.4)^2}{30.4} + \dfrac{(113-121.6)^2}{121.6} = \dfrac{73.96}{30.4} + \dfrac{73.96}{121.6} = 2.433 + 0.608 = 3.041$

Chi-Square Goodness-of-Fit Test for Observed Counts in Variable: Guess

Category	Observed	Historical Counts	Test Proportion	Expected	Contribution to Chi-Sq
Correct	39	0.2	0.2	30.4	2.43289
Incorrect	113	0.8	0.8	121.6	0.60822

N	DF	Chi-Sq	P-Value
152	1	3.04112	0.081

Section 10.2: The Chi-Square Test for Goodness of Fit

10.15 Chi-square goodness-of-fit tests are applicable if the data consist of <u>one categorical variable</u>.

10.17 The answers follow the guidance on page 526.

Step 1: H₀: Humans are like random number generators and produce numbers in equal proportions. Hₐ: Humans are not like random number generators and are not equally likely to pick all the integers.

Step 2: Chi-square test for goodness of fit: *Random Sample* and *Independent Measurements* are met. *Large Sample*: Since the integers are equally likely to be picked, the expected counts are all $(1/5)(38) = 7.6 > 5.$ $\alpha = 0.05$

Step 3: $X^2 = 11.47$, p-value = 0.022

$$X^2 = \dfrac{(3-7.6)^2}{7.6} + \dfrac{(5-7.6)^2}{7.6} + \dfrac{(14-7.6)^2}{7.6} + \dfrac{(11-7.6)^2}{7.6} + \dfrac{(5-7.6)^2}{7.6} = 11.47$$

Chi-Square Goodness-of-Fit Test for Observed Counts in Variable: Times Chosen

Category	Observed	Test Proportion	Expected	Contribution to Chi-Sq
1	3	0.2	7.6	2.78421
2	5	0.2	7.6	0.88947
3	14	0.2	7.6	5.38947
4	11	0.2	7.6	1.52105
5	5	0.2	7.6	0.88947

N	DF	Chi-Sq	P-Value
38	4	11.4737	0.022

Step 4: Reject the H₀ and pick option ii.

10.19 Step 1: H₀: $p_{heads} = 0.50$, Hₐ: $p_{heads} \neq 0.50$, or the coin is biased.

Step 2: Chi-square test for goodness of fit: *Random Sample* and *Independent Measurements* are met. *Large Sample*: The expected counts are both $0.50(50) = 25 > 5.$ $\alpha = 0.05$

Step 3: $X^2 = 8.00$, p-value = 0.005

$$X^2 = \dfrac{(15-25)^2}{25} + \dfrac{(35-25)^2}{25} = 8.00$$

Chi-Square Goodness-of-Fit Test for Observed Counts in Variable: Coin Toss

Category	Observed	Test Proportion	Expected	Contribution to Chi-Sq
1	15	0.5	25	4
2	35	0.5	25	4

N	DF	Chi-Sq	P-Value
50	1	8	0.005

Step 4: Reject H₀. The coin is biased.

10.21 Step 1: H$_0$: All four outcomes are equally likely.
       H$_a$: The four outcomes are not equally likely.

Step 2: Chi-square for goodness of fit; all expected values are 1/4(40) = 10 > 5.

Step 3: Significance level: 0.05, Chi-square = 40.40, p-value= approximately 0.

$$X^2 = \frac{(5-10)^2}{10} + \frac{(1-10)^2}{10} + \frac{(7-10)^2}{10} + \frac{(27-10)^2}{10} = 40.40$$

Observed: Observed
Expected: Expected

N	DF	Chi-Square	P-value
40	3	40.4	<0.0001

Observed #	Expected #
5	10
1	10
7	10
27	10

Step 4: We can reject the null hypothesis. We have shown a significant difference that the wooden dreidel is biased.

10.23 Step 1: H$_0$: The die is fair and produces a proportion of 1/6 in each possible outcome. H$_a$: The die is not fair and does not produce proportions of 1/6 for each possible outcome.

Step 2: Chi-square test for goodness of fit: *Random Sample* and *Independent Measurements* are met. *Large Sample*: The expected counts are all $(1/6)(200) = 20 > 5$. $\alpha = 0.05$

Step 3: $X^2 = 9.20$, p-value = 0.101.

$$X^2 = \frac{(27-20)^2}{20} + \frac{(20-20)^2}{20} + \frac{(22-20)^2}{20} + \frac{(23-20)^2}{20} + \frac{(19-20)^2}{20} + \frac{(9-20)^2}{20} = 9.20$$

Chi-Square Goodness-of-Fit Test for Observed Counts in Variable: Outcome on Die

Category	Observed	Test Proportion	Expected	Contribution to Chi-Sq
1	27	0.166667	20	2.45
2	20	0.166667	20	0.00
3	22	0.166667	20	0.20
4	23	0.166667	20	0.45
5	19	0.166667	20	0.05
6	9	0.166667	20	6.05

N	DF	Chi-Sq	P-Value
120	5	9.2	0.101

Step 4: We cannot reject H$_0$. The die has not been shown to be unfair.

10.25 a. Step 1: H$_0$: $p = 0.20$, H$_a$: $p \neq 0.20$, where p is the population proportion of people who could correctly identify a Stradivarius violin.

Step 2: Chi-square test for goodness of fit: *Random Sample* is not met, but *Random Assignment* is met. Test to see whether the results could easily occur by chance. *Independent Measurements* are met. *Large Sample*: The smallest expected count is $0.2(152) = 30.4 > 5$. $\alpha = 0.05$

Step 3: $X^2 = 3.04$, p-value = 0.081.

$$X^2 = \frac{(39-30.4)^2}{30.4} + \frac{(113-121.6)^2}{121.6} = 3.041$$

```
Chi-Square Goodness-of-Fit Test for Observed Counts in Variable: Guess
                       Historical      Test               Contribution
Category    Observed      Counts  Proportion  Expected       to Chi-Sq
Correct           39         0.2         0.2      30.4         2.43289
Incorrect        113         0.8         0.8     121.6         0.60822
  N  DF  Chi-Sq  P-Value
152   1 3.04112    0.081
```

Step 4: Because the p-value is 0.081, which is more than 0.05, we conclude that the results are not significantly different from guessing, which would produce about 20% correct identifications.

b. Step 1: H_0: $p = 0.20$, H_a: $p > 0.20$, where p is the population proportion of people who could correctly identify a Stradivarius violin.

Step 2: One-proportion z-test, $\alpha = 0.05$

Random Sample: Not met, but *Random Assignment* met, test is results could occur by random chance.

Large Sample: $np_0 = 152(0.20) = 30.4 > 10$ and $n(1 - p_0) = 152(0.80) = 121.6 > 10$

Large Population: There are more than $10 \times 152 = 1520$ professional musicians in the population.

Independence: Assumed.

Step 3: $z = \dfrac{\hat{p} - p_0}{\sqrt{\dfrac{p_0(1 - p_0)}{n}}} = \dfrac{39/152 - 0.20}{\sqrt{\dfrac{0.20(0.80)}{152}}} = 1.74$, p-value $= 0.041$.

Test and CI for One Proportion
```
Test of p = 0.2 vs p > 0.2
                              95% Lower
Sample   X    N   Sample p      Bound   Z-Value  P-Value
1       39  152   0.256579   0.198311      1.74    0.041
Using the normal approximation.
```

Step 4: Because the p-value is 0.041, we can reject H_0 and conclude that there was a significantly higher proportion of correct identification than 20%.

c. The one-sided hypothesis has a p-value that is half that of the two-sided hypothesis, and you can therefore reject H_0 with the one-sided hypothesis.

Section 10.3: Chi-Square Tests for Associations between Categorical Variables

10.27 Independence: one sample.

10.29 Independence: one sample.

10.31 The data are the entire population (not a sample), and therefore there is no need for inference. The data are given as rates (percentages), not frequencies (counts), and there is not enough information for us to convert these percentages to counts.

10.33 Step 1: H_0: The variables fitness app use and gender are independent (not associated),
        H_a: The variables fitness app use and gender are not independent (are associated);

Step 2: Chi-square test of independence, all expected counts are greater than 5 (see table below);

Step 3: Significance level: 0.05, $X^2 = 3.36$, p-value $= 0.067$;

Rows: Use
Columns: None

Cell format
Count
(Expected count)

	Male	Female	Total
Yes	84 (78.32)	268 (273.68)	352
No	9 (14.68)	57 (51.32)	66
Total	93	325	418

Chi-Square test:

Statistic	DF	Value	P-value
Chi-square	1	3.3605955	0.0668

Step 4: Do not reject H_0. Fitness app use has not been shown to be associated with gender.

10.35   Step 1: H_0 Vaccination rates and race are independent (not associated), H_a: Vaccination rates and race are not independent (are associated);

Step 2: Chi-square test of independence, all expected counts are greater than 5 (see table below);

Step 3: Significance level: 0.05, $X^2 = 58.96$, p-value = approximately 0;

Rows: Completed HPV vaccinations
Columns: None

Cell format
Count
(Expected count)

	AAPI	White	Total
Yes	136 (201.54)	1170 (1104.46)	1306
No	216 (150.46)	759 (824.54)	975
Total	352	1929	2281

Chi-Square test:

Statistic	DF	Value	P-value
Chi-square	1	58.96058	<0.0001

Step 4: Reject H_0. Vaccination rates and race are associated.

10.37   a.   Independence: one sample with two variables.

b.   Step 1: H_0: Gender and happiness of marriage are independent. H_a: Gender and happiness of marriage are associated (not independent).

Step 2: Chi-square test of independence: *Random Sample* and *Independent Observations* are met. *Large Sample*: the smallest expected count is $11.88 > 5$. $\alpha = 0.05$ 10.37

Step 3: $X^2 = 10.17$, p-value = 0.006.

$$X^2 = \frac{(278 - 269.22)^2}{269.22} + \frac{(311 - 319.78)^2}{319.78} + \frac{(128 - 128.90)^2}{128.90}$$
$$+ \frac{(154 - 153.10)^2}{153.10} + \frac{(4 - 11.88)^2}{11.88} + \frac{(22 - 14.12)^2}{14.12} = 10.17$$

```
Chi-Square Test: Male, Female
Expected counts are printed below observed counts
         Male   Female   Total
   1      278    311      589
        269.22  319.78
   2      128    154      282
        128.90  153.10
   3        4     22       26
         11.88   14.12
          5.230   4.403
Total    410    487      897
Chi-Sq = 10.173, DF = 2, P-Value = 0.006
```

Step 4: You can reject H_0. Gender and happiness have been shown to be associated.

c. The rate of happiness of marriage has been found to be significantly different for men and women.

10.39 a. The high school graduation rate for not attending preschool is $29/64$, or 45.3%, which is lower than the high school graduation rate of $37/57$, or 64.9%, for attending preschool.

  b. Step 1: H_0: Graduation and preschool are independent. H_a: Graduation and preschool are not independent (they are associated).

   Step 2: Chi-square test of homogeneity: *Random Samples* not met, but *Random Assignment* is met and *Independent Samples and Observations* are met. *Large Sample*: the smallest expected count is $25.91 > 5$, $\alpha = 0.05$

   Step 3: $X^2 = 4.67$, p-value $= 0.031$.

$$X^2 = \frac{(37-31.09)^2}{31.09} + \frac{(29-34.91)^2}{34.91} + \frac{(20-25.91)^2}{25.91} + \frac{(35-29.09)^2}{29.09} = 4.67$$

```
Chi-Square Test: Preschool, No Preschool
Expected counts are printed below observed counts
                         No
           Preschool  Preschool  Total
     1         37        29        66
             31.09      34.91
     2         20        35        55
             25.91      29.09
Total         57        64       121
Chi-Sq = 4.671, DF = 1, P-Value = 0.031
```

Step 4: Reject H_0. Graduation and preschool are associated; can conclude causality, but cannot generalize.

10.41 a. For boys that attended preschool, $16/32$, or 50%, graduated high school, and for those that did not attend preschool, $21/39$, or 53.8%, graduated high school. It is surprising to see that the boys who did not go to preschool had a bit higher graduation rate.

  b. Step 1: H_0: For the boys, graduation and preschool are independent. H_a: For the boys, graduation and preschool are associated.

   Step 2: Chi-square test for homogeneity: *Random Samples* not met, but *Random Assignment* is met and *Independent Samples and Observations* are met. *Large Sample*: the smallest expected count is $15.32 > 5$. $\alpha = 0.05$

   Step 3: $X^2 = 0.104$, p-value $= 0.747$.

$$X^2 = \frac{(16-16.68)^2}{16.68} + \frac{(21-20.32)^2}{20.32} + \frac{(16-15.32)^2}{15.32} + \frac{(18-18.68)^2}{18.68} = 0.105$$

   (0.104 using technology without rounding.)

```
Chi-Square Test: Preschool, No Preschool
Expected counts are printed below observed counts
                              No
           Preschool   Preschool   Total
     1           16          21      37
              16.68       20.32
     2           16          18      34
              15.32       18.68
Total           32          39      71
Chi-Sq = 0.104, DF = 1, P-Value = 0.747
```

Step 4: Do not reject H_0. For the boys, there is no evidence that attending preschool is associated with graduating from high school.

c. The results do not generalize to other groups of boys and girls, but what evidence we have suggests that although preschool might be effective for girls, it may not be for boys, at least with regard to graduation from high school.

10.43 a.

Supports Marijuana Legalization		
Generation	Yes	No
Millennial	140	60
GenX	132	68

b. Step 1: H_0: Support of marijuana legalization and generation are independent (not associated), H_a: Support of marijuana legalization and generation are not independent (are associated).

Step 2: Chi-square test of independence, all expected counts greater than 5 (see table below).

Step 3: Significance level: 0.05, $X^2 = 0.74$, p-value = 0.391

Contingency table results:

Rows: Generation
Columns: None

Cell format
Count
(Expected count)

	Yes	No	Total
Millennial	140	60	200
	(136)	(64)	
GenX	132	68	200
	(136)	(64)	
Total	272	128	400

Chi-Square test:

Statistic	DF	Value	P-value
Chi-square	1	0.73529412	0.3912

Step 4: Do not reject H_0. Support for marijuana legalization and generation have not been shown to be associated.

c. No. Support of marijuana legalization is not significantly different among these two generations.

10.45 a. Tranexamic: $383/1161 = 33.0\%$, Placebo: $419/1164 = 36.0\%$

b.

Adverse Outcome	Drug	Placebo	Total
Yes	383	419	802
No	778	745	1523
Total	1161	1164	2325

c. Step 1: H_0: Treatment and adverse outcome are independent (not associated), H_a: Treatment and adverse outcome are not independent (are associated);

Step 2: Chi-square test of independence, all expected counts greater than 5 (see table below);

Step 3: Significance level: 0.05, $X^2 = 2.33$, p-value = 0.127;

Contingency table results:

Rows: Adverse Outcome
Columns: None

Cell format
Count (Expected count)

	Drug	Placebo	Total
Yes	383 (400.48)	419 (401.52)	802
No	778 (760.52)	745 (762.48)	1523
Total	1161	1164	2325

Chi-Square test:

Statistic	DF	Value	P-value
Chi-square	1	2.3271291	0.1271

Step 4: Do not reject H_0. Treatment and adverse outcome have not been shown to be associated.

10.47　Step 1: H_0: Political party affiliation and education are independent (not associated), H_a: Political party affiliation and education are not independent (are associated);

Step 2: Chi-square test of independence, all expected counts greater than 5 (see table below);

Step 3: Significance level: 0.05, $X^2 = 14.70$, p-value = 0.002;

Contingency table results:

Rows: Educational Attainment
Columns: None

Cell format
Count (Expected count)

	Democrat/Lean Democrat	Republican/Lean Republican	Total
High school or less	144 (160.48)	150 (133.52)	294
Some college	132 (140.83)	126 (117.17)	258
College graduate	135 (127.18)	98 (105.82)	233
Postgraduate degree	95 (77.51)	47 (64.49)	142
Total	506	421	927

Chi-Square test:

Statistic	DF	Value	P-value
Chi-square	3	14.692443	0.0021

Step 4: Reject H_0: Political party affiliation and education are associated.

Section 10.4: Hypothesis Tests When Sample Sizes Are Small

10.49 a.

	Alcohol Use		
Age	None	1–9 days	10+ days
18–20	182	100	31
21–24	142	109	39
25–29	49	41	7
30+	76	32	10

Step 1: H_0: Age group and alcohol use are independent (not associated), H_a: Age group and alcohol use are not independent (are associated);

Step 2: Chi-square test for independence, all expected counts greater than 5 (see table below);

Step 3: Significance level: 0.05, $X^2 = 13.60$, p-value = 0.034;

Contingency table results:

Rows: Age
Columns: None

Cell format
Count
(Expected count)

	None	1-9 days	10+ days	Total
18-20	182 (171.81)	100 (107.9)	31 (33.29)	313
21-24	142 (159.18)	109 (99.98)	39 (30.84)	290
25-29	49 (53.24)	41 (33.44)	7 (10.32)	97
30+	76 (64.77)	32 (40.68)	10 (12.55)	118
Total	449	282	87	818

Chi-Square test:

Statistic	DF	Value	P-value
Chi-square	6	13.598074	0.0345

Step 4: Reject H_0; There is an association between age group and alcohol use.

10.51 a. The smallest expected count is $(54)(879)/1960 = 24.22$ from male and other party, so the original table could have been used.

   b.

	Male	**Female**
Dem	$142 + 136 = 278$	$214 + 207 = 421$
Rep	$115 + 83 = 198$	$135 + 109 = 244$
Other	$127 + 147 + 93 + 36 = 403$	$108 + 226 + 64 + 18 = 416$

   c. Women: $421/1081 = 38.9\%$ Democrats. Men: $278/879 = 31.6\%$ Democrats. Thus these women are more likely to be Democrats than these men.

   d. Step 1: H_0: Gender and political party are independent. H_a: Gender and political party are not independent.

   Step 2: Chi-square test of independence: *Random Sample* and *Independent Observations* must be assumed. *Large Sample*: the smallest expected count is $198.22 > 5$. $\alpha = 0.05$

Step 3: $X^2 = 13.57$, p-value = 0.001.

$$X^2 = \frac{(278-313.48)^2}{313.48} + \frac{(421-385.52)^2}{385.52} + \frac{(198-198.22)^2}{198.22}$$

$$+ \frac{(244-243.78)^2}{243.78} + \frac{(403-367.30)^2}{367.30} + \frac{(416-451.70)^2}{451.70} = 13.57$$

Chi-Square Test: Male, Female
```
Expected counts are printed below observed counts
        Male   Female  Total
  1      278      421    699
       313.48   385.52
  2      198      244    442
       198.22   243.78
  3      403      416    819
       367.30   451.70
Total    879     1081   1960
Chi-Sq = 13.574, DF = 2, P-Value = 0.001
```

Step 4: Reject H_0. The difference is significant. Gender and political party are not independent.

10.53 a. You would get a smaller p-value, because it is more extreme in the direction of the antivenom working.

   b. You would get a larger p-value, because the results are less extreme.

   c. The p-value for the test in part a is 0.0002 (both one-tailed and two-tailed); yes, it is smaller.

Tabulated statistics: Outcome, Treatment
```
Rows: Outcome    Columns: Treatment
        Antivenom  Placebo  All
Bad            0        7     7
Good           8        0     8
All            8        7    15
Fisher's exact test: P-Value =  0.0001554
```

The p-value for the test in part b is 0.1002 (one-tailed) and 0.1319 (two-tailed); yes, it is larger.

Tabulated statistics: Outcome, Treatment
```
Rows: Outcome    Columns: Treatment
        Antivenom  Placebo  All
Bad            6        2     8
Good           2        5     7
All            8        7    15
Fisher's exact test: P-Value =  0.131935
```

10.55 a. Consume peanuts: 5/272 = 1.8%, Avoid peanuts: 37/270 = 13.7%

   b. Step 1: H_0: Treatment group and peanut allergy are independent (not associated); H_a: Treatment group and peanut allergy are not independent (are associated);

   Step 2: Chi-square test for independence, all expected counts greater than 5(see table below);

Step 3: Significance level: 0.05, $X^2 = 26.69$, p-value = approximately 0;

Contingency table results:

Rows: Peanut Allergy at age 60mos
Columns: None

Cell format
Count
(Expected count)

	Consume Peanuts	Avoid Peanuts	Total
Yes	5 (21.08)	37 (20.92)	42
No	267 (250.92)	233 (249.08)	500
Total	272	270	542

Chi-Square test:

Statistic	DF	Value	P-value
Chi-square	1	26.685936	<0.0001

Step 4: Reject H_0: There is an association between treatment group and peanut allergy.

c. Fisher's Exact Test p-value < 0.0001

Contingency table results:

Rows: Peanut Allergy at age 60mos
Columns: None

Cell format
Count
(Expected count)

	Consume Peanuts	Avoid Peanuts	Total
Yes	5 (21.08)	37 (20.92)	42
No	267 (250.92)	233 (249.08)	500
Total	272	270	542

Fisher's exact test:

P-value = <0.0001

d. Chi-square p-value = approximately 0. Fisher's Exact Test p-value < 0.0001. Fisher's Exact test is correct.

Chapter Review Exercises

10.57 Chi-square goodness-of-fit test

10.59 Chi-square test of independence (one sample, two variables)

10.61 Chi-square test of independence, one sample

10.63 No test because the data are a population, not a sample

10.65 a. 31/65, or 47.7%, of those in the control group were arrested, and 8/58, or 13.8%, of those who attended preschool were arrested. Thus there was a lower rate of arrest for those who went to preschool.

   b.

	Preschool	No Preschool
Arrest	8	31
No Arrest	50	34

Step 1: H_0: The treatment and arrest rate are independent. H_a: The treatment and arrest rate are associated.

Step 2: Chi-square for homogeneity: *Random Sample* not met, but *Random Assignment* is met and *Independent Samples and Observations* is assumed. *Large Sample*: the smallest expected count is $18.39 > 5$. $\alpha = 0.05$

Step 3: $X^2 = 16.27$, p-value = 0.000055 or p-value < 0.001.

$$X^2 = \frac{(8-18.39)^2}{18.39} + \frac{(31-20.61)^2}{20.61} + \frac{(50-39.61)^2}{39.61} + \frac{(34-44.39)^2}{44.39} = 16.27$$

```
Chi-Square Test: Preschool, No Preschool
Expected counts are printed below observed counts
                        No
          Preschool  Preschool  Total
    1           8        31       39
            18.39     20.61
    2          50        34       84
            39.61     44.39
Total          58        65      123
Chi-Sq = 16.266, DF = 1, P-Value = 0.000
```

Step 4: We can reject the hypothesis of no association at the 0.05 level. We conclude that preschool attendance affects the arrest rate, but cannot generalize the results.

c.   Step 1: H_0: $p_{\text{Preschool}} = p_{\text{No Preschool}}$, H_a: $p_{\text{Preschool}} < p_{\text{No Preschool}}$, where p is the rate of arrest.

Step 2: Two-proportion z-test, $\alpha = 0.05$

$$\hat{p} = \frac{8+31}{58+65} = \frac{13}{41}, \quad n_{\text{Preschool}}\hat{p} = 58\left(\frac{13}{41}\right) = 18.4 > 10, \quad n_{\text{Preschool}}(1-\hat{p}) = 58\left(\frac{28}{41}\right) = 39.6 > 10,$$

$$n_{\text{No Preschool}}\hat{p} = 65\left(\frac{13}{41}\right) = 20.6 > 10, \quad n_{\text{No Preschool}}(1-\hat{p}) = 65\left(\frac{28}{41}\right) = 44.4 > 10$$

Random Assignment, Independence, both within samples and between samples, and *Large Populations* are met.

Step 3: $z = \dfrac{\left(\hat{p}_{\text{Preschool}} - \hat{p}_{\text{No Preschool}} - 0\right)}{\sqrt{\hat{p}(1-\hat{p})\left(\dfrac{1}{n_1} + \dfrac{1}{n_2}\right)}} = \dfrac{(8/58 - 31/65 - 0)}{\sqrt{\dfrac{13}{41}\left(1-\dfrac{13}{41}\right)\left(\dfrac{1}{58} + \dfrac{1}{65}\right)}} = -4.03,$

(4.03 if proportions are reversed), p-value = 0.000028 or p-value < 0.001.

```
Test and CI for Two Proportions
Sample    X    N   Sample p
1         8   58   0.137931
2        31   65   0.476923
Difference = p (1) - p (2)
Estimate for difference:  -0.338992
95% upper bound for difference:  -0.212776
Test for difference = 0 (vs < 0):  Z = -4.03  P-Value = 0.000
```

Step 4: Reject the null hypothesis. Preschool lowers the rate of arrest, but we cannot generalize.

d.   The z-test enables us to test the alternative hypothesis that preschool attendance lowers the risk of later arrest. The Chi-square test allows for testing for some sort of association, but we can't specify whether it is a positive or a negative association. Note that the p-value for the one-sided hypothesis with the z-test is half the p-value for the two-sided hypothesis with the Chi-square test.

10.67   It would not be appropriate because the data are percentages (not counts), and we cannot convert them to counts given the information in the problem.

10.69   a.   *Men*: $120 / 500 = 24.0\%$ and *Women*: $182 / 519 = 35.1\%$

b. Step 1: H$_0$: Gender and experience of sexual harassment in the workplace are independent; H$_a$: Gender and experience of sexual harassment in the workplace are not independent;

Step 2: Chi-square test of independence; all expected counts greater than 5 (see table below);

Step 3: Significance level: 0.05, $X^2 = 14.96$, p-value = 0.0001;

Contingency table results:

Rows: Gender
Columns: None

Cell format
Count (Expected count)

	Yes	No	Total
Men	120 (148.18)	380 (351.82)	500
Women	182 (153.82)	337 (365.18)	519
Total	302	717	1019

Chi-Square test:

Statistic	DF	Value	P-value
Chi-square	1	14.958209	0.0001

Step 4: Reject H$_0$: Gender and experience of sexual harassment

10.71 a. Step 1: H$_0$: The presence or absence of robots is independent of grouping.
H$_a$: The presence or absence of robots is not independent of grouping.

Step 2: Chi-square test for homogeneity: *Random Samples* and *Independent Samples and Observations* are given. *Large Samples*: two expected counts are 5, which is on the low side for this approximation. $\alpha = 0.05$

Step 3: $X^2 = 4.32$, p-value = 0.038.

$$X^2 = \frac{(22-25)^2}{25} + \frac{(28-25)^2}{25} + \frac{(8-5)^2}{5} + \frac{(2-5)^2}{5} = 4.32$$

```
Chi-Square Test: Cockroaches Only, Robots Also
Expected counts are printed below observed counts
          Cockroaches  Robots
                 Only    Also  Total
       1           22      28     50
                25.00   25.00
       2            8       2     10
                 5.00    5.00
   Total          30      30     60
   Chi-Sq = 4.320, DF = 1, P-Value = 0.038
```

Step 4: Reject H$_0$. The proportions are not the same. The robots have a significant effect.

b. With Fisher's Exact Test, using a two-sided alternative, the p-value is 0.080. You cannot reject H$_0$. The robots do not have a significant effect.

```
Tabulated statistics: Shelter Used, Group
Rows: Shelter Used   Columns: Group
          Roaches
            and  Roaches
         Robots    Only  All
   Both        2       8   10
   One        28      22   50
   All        30      30   60
   Cell Contents:    Count
   Fisher's exact test: P-Value =  0.0797220
```

c. For chi-square, the p-value was 0.038, and with Fisher's Exact Test, the p-value was 0.080. Fisher's Exact Test is accurate, and the chi-square test is a *Large Sample* approximation. The approximation is not very good with two expected counts of 5, and that is why the two p-values are so different. The p-value of 0.080 is the accurate value.

10.73 a. 19/67, or 28.4%, of the minority defendants were convicted.

b. 38/118, or 32.2%, of the white defendants were convicted.

c. Step 1: H_0: Race and conviction are independent. H_a: Race and conviction are not independent.

Step 2: Chi-square test of independence: *Random Sample* and *Independent Observations* are met. *Large Sample*: the smallest expected count is 20.6 > 5. $\alpha = 0.05$

Step 3: $X^2 = 0.30$, p-value = 0.586.

$$X^2 = \frac{(48-46.36)^2}{46.36} + \frac{(80-81.64)^2}{81.64} + \frac{(19-20.64)^2}{20.64} + \frac{(38-36.36)^2}{36.36} = 0.30$$

```
Chi-Square Test: Non-white, White
Expected counts are printed below observed counts
         Non-white  White  Total
   1           48     80     128
             46.36  81.64
   2           19     38      57
             20.64  36.36
Total          67    118     185
Chi-Sq = 0.296, DF = 1, P-Value = 0.586
```

Step 4: Do not reject H_0. Race and conviction have not been shown to be associated.

Chapter 11: Multiple Comparisons and Analysis of Variance

Answers may vary slightly due to rounding and type of technology used. All t-statistics are reported as positive values, although the sign of your t-statistics (positive or negative) must be consistent with which group you chose for group 1.

Section 11.1: Multiple Comparisons

All calculations for this section were done without assuming equal variances.

11.1   a.   ANOVA

        b.   Two-sample *t*-test

11.3   a. $5(5-1)/2 = 20/2 = 10$,   AB, AC, AD, AE, BC, BD, BE, CD, CE, DE

        b. $0.05/10 = 0.0050$

11.5

Comparison	*t*-statistic	p-value	Conclusion
Seattle - SF	5.17	0.0003	Significantly different
Seattle – Santa Monica	-3.36	0.006	Significantly different
SF – Santa Monica	2.27	0.043	Not significantly different

11.7   a.   With three groups there are three possible comparisons, so the corrected value for alpha is $\alpha = 0.05/3$, or about 0.0167.

        b.   Sample means: LA - \$3.46, Chicago - \$3.13 Boston - \$2.94. The closest means are Chicago and Boston.

        c.

Comparison	*t*-statistic	p-value	Conclusion
LA - Chicago	20.05	<0.0001	Significantly different
LA - Boston	25.68	<0.0001	Significantly different
Chicago - Boston	7.78	<0.0001	Significantly different

11.9   a.   3

        b. $0.05/3$, or about 0.0167

        c. $1 - 0.0167 = 98.33\%$

11.11

Comparison	Confidence Interval	Conclusion
LA - Chicago	(0.282, 0.378)	Does not capture 0, reject equality.
LA - Boston	(0.463, 0.585)	Does not captures 0, reject equality.
Chicago - Boston	(0.126, 0.262)	Does not captures 0, reject equality.

These conclusions are the same as those for exercise 11.7. All of the means are significantly different from each other. These are the same conclusions reached in exercise 11.7.

11.13   a.   Sample Means: Shortstop: 50.6, Left Field: 51.3, First Base: 52.3

         b.   3 comparisons, Bonferroni-corrected significance level: $0.05/3 = 0.0167$

         c.

Comparison	*t*-statistic	p-value	Conclusion
SS - LF	-0.16	0.875	Not Significantly different
SS - FB	-0.47	0.648	Not Significantly different
LF - FB	-0.23	0.820	Not Significantly different

Section 11.2: The Analysis of Variance

11.15 The *F*-value of 9.38 goes with A, B, and C. The *F*-value of 150.00 goes with L, M, and N. The reason for the difference is that the variation between groups (the separation between means) is larger for L, M, and N, relative to the variation within groups (which is the same in all groups).

$$F = \frac{\text{Variation between Groups}}{\text{Variation within Groups}}$$

11.17 a. H₀: The population means are all equal, or marital status and cholesterol levels are not associated.
H_a: At least population one mean is different from another, or marital status and cholesterol levels are associated.

b. $F = 9.14$

c. Largest mean cholesterol: divorced. Smallest mean cholesterol: never married.

d. This was an observational study, from which you cannot conclude causality. One possible confounder is age. For example, the never married may tend to be young, and youth may cause the low cholesterol.

11.19 a. SS Error = 7790.5 – 893.5 = 6897

b. MS Error = 6897/106 = 65.066, which rounded is 65.1.

c. $297.8 / 65.066 = 4.5769$, which rounded is 4.58.

d. When MS factor is more than MS Error, the *F*-value will be more than 1.

11.21 a. The highest sample mean was for the freshmen, and the lowest was for the seniors.

b. μ is the population mean number of hours of school work per week. H₀: $\mu_1 = \mu_2 = \mu_3 = \mu_4$,
H_a: At least population one mean is different from another, or class has an effect on school work.

c. $F = 4.58$

d. No. There was no random assignment. There could be confounding factors, such as age, hours of work for money, or living situation.

Section 11.3: The ANOVA Test

11.23

Price	Code
2.75	1
2.74	1
2.72	1
2.73	1
2.75	1
2.70	2
2.68	2
2.66	2
2.71	2
2.74	2
2.85	3
2.83	3
2.84	3
3.01	3
2.97	3

11.25 p-value = 0.005. Reject H₀. The mean amount of school work does vary by class.

11.27 The pulse rates are not in three independent groups, so the condition of independent groups fails.

11.29 Do not use ANOVA, because the standard deviations are too different. The ratio of the largest to the smallest is $34.02/14.36$, or about 2.37, which is larger than 2.

11.31 Step 1: H_0: $\mu_{\text{Seattle}} = \mu_{\text{SF}} = \mu_{\text{Santa Monica}}$; H_a: At least one population mean is different from another;

Step 2: One-way ANOVA. Distributions satisfy ANOVA conditions;

Step 3: Significance level: 0.05, $F = 14.16$, p-value = 0.0001;

Analysis of Variance results:

Data stored in separate columns.

Column statistics

Column	n	Mean	Std. Dev.	Std. Error
Seattle	8	1723.5	408.84401	144.54819
San Fransisco	8	3275	744.19659	263.11323
Santa Monica	8	2535.125	547.49154	193.56749

ANOVA table

Source	DF	SS	MS	F-Stat	P-value
Columns	2	9635473.1	4817736.5	14.159694	0.0001
Error	21	7145102.9	340242.99		
Total	23	16780576			

Step 4: Reject H_0. The mean rents for one-bedroom apartments in these cities are not equal.

11.33 Step 1: H_0: $\mu_{\text{IL}} = \mu_{\text{LA}} = \mu_{\text{P}} = \mu_{\text{S}}$, H_a: At least population one mean is different from another.

Step 2: Choose one-way ANOVA. *Random Sample and Independent Measurements, Independent Groups,* and *Normal Distribution or Large Sample* are assumed.
Same Variance: $4369/3219 = 1.36 < 2$. $\alpha = 0.05$

Step 3: $F = 39.02$, p-value < 0.001.

Step 4: Reject H_0. Type of school does affect median starting salary.

11.35 a. The medians and interquartile ranges are all similar, although the median for the athletes was a little larger (slower!) than the others, and the interquartile range was a bit larger for the moderate group. Also, the shapes are not strongly skewed. There are no potential outliers.

b. Step 1: H_0: $\mu_{\text{NotAth}} = \mu_{\text{Mod}} = \mu_{\text{Athletic}}$, H_a: At least population one mean is different from another.

Step 2: Choose one-way ANOVA: *Random Sample and Independent Measurements, Independent Groups,* and *Normal Distribution or Large Sample* are assumed.
Same Variance: $6.866/5.443 = 1.26 < 2$. $\alpha = 0.05$

Step 3: $F = 0.10$, p-value $= 0.903$.

```
One-way ANOVA: NotAthletic, Medium, Athletic
Source   DF      SS     MS      F      P
Factor    2     8.2    4.1   0.10   0.903
Error    26  1042.3   40.1
Total    28  1050.5
S = 6.331    R-Sq = 0.78%    R-Sq(adj) = 0.00%
                             Individual 95% CIs For Mean Based on
                             Pooled StDev
Level          N    Mean   StDev  --------+---------+---------+---------+
NotAthletic    7  20.843   6.866        (---------------*----------------)
Medium        12  20.142   6.693          (-----------*------------)
Athletic      10  19.440   5.443  (-------------*-------------)
                                   --------+---------+---------+---------+
                                       18.0      21.0      24.0      27.0
Pooled StDev = 6.331
```

Step 4: Do not reject Ho. The sample means were not significantly different, and we conclude that we cannot reject the hypothesis that the population means are the same.

11.37   Step 1: Ho: $\mu_{front} = \mu_{middle} = \mu_{back}$, Ha: At least one group mean differs from another.

Step 2: Choose one-way ANOVA: *Random Sample and Independent Measurements, Independent Groups,* and *Normal Distribution or Large Sample* are assumed.
Same Variance: $0.4637 / 0.2574 = 1.80 < 2$. $\alpha = 0.05$

Step 3: $F = 7.50$, p-value $= 0.003$.

```
One-way ANOVA: Front, Middle, Back
Source   DF    SS      MS      F      P
Factor    2   2.350   1.175   7.50   0.003
Error    27   4.229   0.157
Total    29   6.580
S = 0.3958   R-Sq = 35.72%   R-Sq(adj) = 30.96%
                             Individual 95% CIs For Mean Based on
                             Pooled StDev
Level     N    Mean   StDev  -----+---------+---------+---------+----
Front    10   3.4486  0.4344                        (--------*--------)
Middle   10   2.9100  0.4637           (--------*--------)
Back     10   2.8119  0.2574  (--------*-------)
                             -----+---------+---------+---------+----
                             2.70      3.00      3.30      3.60
Pooled StDev = 0.3958
```

Step 4: Reject Ho. The mean GPA is not the same for all rows.

11.39   Step 1: Ho: $\mu_{excellent} = \mu_{fair} = \mu_{good} = \mu_{poor}$, (suggesting health status and hours of sleep are independent),
Ha: At least population one mean is different from another.

Step 2: Choose one-way ANOVA: *Random Sample and Independent Measurements, Independent Groups,* and *Normal Distribution or Large Sample* are assumed.
Same Variance: $1.786 / 0.984 = 1.998 < 2$. $\alpha = 0.05$

Step 3: $F = 4.44$, p-value $= 0.0053$.

Step 4: Reject Ho. The means are not all equal. Health status and hours of sleep are not independent at this company.

Section 11.4: Post-Hoc Procedures

All calculations for post-hoc tests were done with pooled variances.

11.41   The UT-CA interval does not contain 0. Since both limits are negative, CA has longer travel times than UT. The NY-UT interval also does not contain 0. Since both limits are positive, NY has longer travel times than UT. UT has the shortest travel times of these 3 states.

11.43   Step 1: Ho: $\mu_{Front} = \mu_{Middle} = \mu_{Back}$, Ha: At least population one mean is different from another.

Step 2: Choose one-way ANOVA: *Random Sample and Independent Measurements, Independent Groups,* and *Normal Distribution or Large Sample* are assumed.
Same Variance: $0.4637 / 0.2574 = 1.80 < 2$. $\alpha = 0.05$

Step 3: $F = 7.50$, p-value $= 0.003$.

```
One-way ANOVA: Front, Middle, Back
Source   DF    SS      MS      F      P
Factor    2   2.350   1.175   7.50   0.003
Error    27   4.229   0.157
Total    29   6.580
S = 0.3958   R-Sq = 35.72%   R-Sq(adj) = 30.96%
                             Individual 95% CIs For Mean Based on
                             Pooled StDev
```

```
Level    N    Mean    StDev   -----+---------+---------+---------+----
Front   10   3.4486   0.4344                       (--------*--------)
Middle  10   2.9100   0.4637        (--------*--------)
Back    10   2.8119   0.2574   (--------*-------)
                                -----+---------+---------+---------+----
                                   2.70      3.00      3.30      3.60

Pooled StDev = 0.3958
```

Step 4: Reject H_0. GPA and row level appear to be associated.

Post-hoc procedures:

Comparison	Tukey CI	TI-84 CI (Bonferroni)	Reject H_0?
Middle – Front	$(-0.98, -0.10)$	$(-1.07, -0.01)$	Does not capture 0: Yes
Back – Front	$(-1.08, -0.20)$	$(-1.06, -0.22)$	Does not capture 0: Yes
Back – Middle	$(-0.54, 0.34)$	$(-0.54, 0.34)$	Captures 0: No

Back   Middle   Front

```
Tukey 95% Simultaneous Confidence Intervals
All Pairwise Comparisons
Individual confidence level = 98.04%
Front subtracted from:
          Lower    Center   Upper    --+---------+---------+---------+-------
Middle  -0.9779  -0.5386  -0.0993      (--------*--------)
Back    -1.0760  -0.6367  -0.1974    (--------*--------)
                                      --+---------+---------+---------+-------
                                      -1.00     -0.50     0.00      0.50

Middle subtracted from:
          Lower    Center   Upper    --+---------+---------+---------+-------
Back    -0.5374  -0.0981   0.3412              (--------*--------)
                                      --+---------+---------+---------+-------
                                      -1.00     -0.50     0.00      0.50
```

Those who sit in the front row tend to have higher GPAs, on average, then those who sit either in the back or in the middle. There is no distinguishable difference in mean GPAs between those in the back and those in the middle.

11.45 From Exercise 11.35, $F = 0.10$, p-value $= 0.903$. We cannot reject the null hypothesis of no differences in population mean reaction distances, so you should not do post-hoc tests. In other words, the sample means are not significantly different.

11.47 Since the confidence interval limits of SF-Seattle are both positive, SF is more expensive than Seattle. Since the confidence interval limits of Santa Monica-Seattle are both positive, Santa Monica is also more expensive than Seattle. Since the confidence interval limits of Santa Monica-SF are both negative, SF is more expensive than Santa Monica. The cities (in order of least expensive to most expensive): Seattle, Santa Monica, San Francisco.

11.49 Step 1: H_0: $\mu_{Democrat} = \mu_{Independent} = \mu_{Other} = \mu_{Republican}$, H_a: At least population one mean is different from another.

Step 2: Choose one-way ANOVA: *Random Sample and Independent Measurements, Independent Groups,* and *Normal Distribution or Large Sample* are assumed.
Same Variance: $36.3353 / 31.9610 = 1.14 < 2$. $\alpha = 0.05$

Step 3: $F = 5.98$, p-value $= 0.0005$.

Step 4: Reject H_0. Level of concern and party affiliation are not independent for StatCrunch users.

Means, from smallest to largest: Republican   Other   Independent   Democrat

Republicans are significantly less concerned than all the other parties. There are no other significant differences.

Chapter Review Exercises

11.51 No, since multiple *t*-tests were performed. There are a total of 7 groups and therefore 21 possible comparisons. The Bonferroni-corrected significance level would be $\alpha = 0.05/21$, or about 0.0024. The p-value of 0.012 is not less than 0.0024.

11.53 Because all the intervals capture 0, we have not found any significant differences in the mean length of time since contacting their mother for the three groups of professors.

Post-hoc procedures:

Comparison	Bonferroni Confidence Interval	Conclusion
Eth – Phil	$(-5.21, 17.81)$	Captures 0: Not Different
Eth – Other	$(-1.28, 20.35)$	Captures 0: Not Different
Phil – Other	$(-1.91, 8.37)$	Captures 0: Not Different

```
Two-Sample T-Test and CI: Eth, Phil
Difference = mu (Eth) - mu (Phil)
Estimate for difference:   6.30
98.33% CI for difference:   (-5.21, 17.81)
Two-Sample T-Test and CI: Eth, Other
Difference = mu (Eth) - mu (Other)
Estimate for difference:   9.53
98.33% CI for difference:   (-1.28, 20.35)
Two-Sample T-Test and CI: Phil, Other
Difference = mu (Phil) - mu (Other)
Estimate for difference:   3.23
98.33% CI for difference:   (-1.91, 8.37)
```

11.55 Step 1: H_0: $\mu_{EthChar} = \mu_{PhilChar} = \mu_{OtherChar}$, H_a: At least population one mean is different from another.

Step 2: Choose one-way ANOVA: *Random Sample and Independent Measurements, Independent Groups*, and *Normal Distribution or Large Sample* are assumed.
Same Variance: $4.821/4.259 = 1.13 < 2$. $\alpha = 0.05$

Step 3: $F = 7.33$, p-value $= 0.001$.

Step 4: Reject H_0. There are significant differences in levels of charitable giving between the groups.

Means, from smallest to largest: PhilChar   OtherChar   EthChar

Philosophy professors report giving a significantly lower percentage to charity than ethicists and professors in other fields. There is no significant difference between ethicists and other professors.

11.57 The p-values are the same. The *F*-statistic is the square of the *t*-statistic. You get the same conclusion either way: Do not reject the null hypothesis of equal population mean triglyceride levels for men and women.

Chapter 12: Experimental Design: Controlling Variation

Answers may vary slightly due to type of technology or rounding.

Section 12.1: Variation Out of Control

12.1   Headline A: Cause and effect

Headline B: No cause and effect

12.3   This was an observational study. Researchers could not assign subjects to smoke marijuana regularly.

12.5   Regular marijuana use leads to increased likelihood of bone fractures (causality). Regular marijuana use associated with increased bone fracture susceptibility (no causality). The second is correct because we cannot infer cause and effect from an observational study.

12.7   a. clopidogrel and aspirin: 121/2432 = 5.0%, placebo and aspirin 159/ 2449 = 6.5%.

b. controlled experiment (random assignment)

c. The treatment variable is the drug (clopidogrel); the response variable is the percentage of subjects who had another stroke.

d. A combination of clopidogrel and aspirin lowers the risk of recurrent stroke in stroke patients compared with treatment with aspirin alone.

12.9   a. The treatment variable is the drug (upadacitinib); the response variable is improvement in rheumatoid arthritis symptoms.

b. The drug upadacitinib causes improvement in symptoms among rheumatoid arthritis patients.

12.11   This was an observational study because subjects were not assigned to exercise groups.

12.13   a. Two-sample proportion test; $z = -1.03$, p-value = 0.301.

Two sample proportion summary hypothesis test:

p_1 : proportion of successes for population 1
p_2 : proportion of successes for population 2
$p_1 - p_2$: Difference in proportions
$H_0 : p_1 - p_2 = 0$
$H_A : p_1 - p_2 \neq 0$

Hypothesis test results:

Difference	Count1	Total1	Count2	Total2	Sample Diff.	Std. Err.	Z-Stat	P-value
$p_1 - p_2$	84	100	89	100	-0.05	0.048327011	-1.0346181	0.3008

Do not reject H_0. We cannot conclude the proportion of men and women

b. Men: .84(2291) = 1924; Women: 0.89(2282) = 2031

Two-sample proportion test, $z = -4.96$, p-value = approximately 0.

Two sample proportion summary hypothesis test:

p_1 : proportion of successes for population 1
p_2 : proportion of successes for population 2
$p_1 - p_2$: Difference in proportions
$H_0 : p_1 - p_2 = 0$
$H_A : p_1 - p_2 \neq 0$

Hypothesis test results:

Difference	Count1	Total1	Count2	Total2	Sample Diff.	Std. Err.	Z-Stat	P-value
$p_1 - p_2$	1924	2291	2031	2282	-0.05020082	0.010111056	-4.9649433	<0.0001

Reject H_0. The proportion of men and women who agree with this statement are different.

c. The larger sample size gave us more power, leading us to reject the null hypothesis.

12.15 a. The first will have more variability because it draws from both men and women.

b. The second study is drawing a sample from a population with less variability and so it will have more power.

12.17 This is not an appropriate use of blocking. In this design, the researchers randomized the blocks, not the subjects. Randomization should happen within blocks. For this study, each patient in a block should be assigned a number, and these numbers should be put into a bowl, mixed up. There will be four bowls, one for each age group. From each bowl (each block), half of the subjects' numbers are chosen, and these people will receive the treatment. The others receive a placebo.

12.19 a. The treatment variable is the speed skating suit; the response variable is the race speed.

b. Randomly assign each skater to race either wearing a Mach 39 suit or an H1 suit. To do this, in a bag put 10 tickets that say "Mach 39" and 10 tickets that say "H1." Each skater chooses a ticket and will use that type of suit. Record the race times.

c. Block on whether the skater is an Olympic skater or not. Randomly assign half the Olympic skaters to the Mach 39 suits by putting in a bag 5 tickets that say "Mach 39" and five tickets that say "H1." Each Olympic skater chooses a ticket and will use that type of suit. Do the same process with the recreational speed skaters, randomly assigning half to use the Mach 39 suit and half to use the H1 suit. The blocked design will prevent having uneven groups with more Olympic skaters in one group than The other.

d. Have each skater race with a Mach 39 suit and also with an H1 suit. Use a paired *t*-test for the comparison. Randomly assign half of all skaters to wear the Mach 39 first and the other half to wear the H1 first. (Otherwise some skaters might skate more slowly in their second race because they were tired from their first race)

12.21 a. The treatment variable records whether subjects get aspirin or placebo. The response variable is whether the person has a heart attack.

b. You could put 100 slips of paper marked A and 100 slips of paper marked B in a bag. Each person would draw out a slip of paper. Those who got A would get the aspirin, and those who got B would get the placebo. Then observe the subjects for a given time interval to see whether they have a heart attack, and compare the percentages with heart attacks for the two groups.

c. Randomly assign half of the men and half of the women to use the aspirin, and assign the rest to use a placebo. You could use two separate bags, one for the men and one for the women. Each bag would have 100 slips of paper: 50 marked A and 50 marked B. Each woman draws randomly from the women's bag, and each man draws randomly from the men's bag. Then observe the subjects for a given time interval to see whether they have a heart attack, and determine the percentage of aspirin-taking men who had a heart attack, the percentage of placebo-taking men who had a heart attack, the percentage of aspirin-taking women who had a heart attack, and the percentage of placebo-taking women who had a heart attack.

d. The blocked design improves statistical power, in part by preventing an uneven distribution of men and women in the two groups. With the blocked design, we have a higher probability of determining whether aspirin reduces the risk of heart attack, if it actually does so.

12.23 a. This is a controlled experiment (used random assignment).

b. Treatment group: 10/33 = 30.3%, Control group: 2/34 = 5.9%.

c. (i) Since the p-value is less than 0.05 the null hypothesis is rejected. Researchers have shown that dietary improvement may be an effective treatment strategy for patients with moderate to severe depression.

12.25 a. The mean for white paper was smaller (the means were 74.3 and 76.3). This contradicts the idea that reading material written on colored paper is easier to read.

b. Step 1: H$_0$: $\mu_{salmon} = \mu_{white}$, H$_a$: $\mu_{salmon} < \mu_{white}$, where μ is the population mean reading time.

Step 2: Two-sample *t*-test: *Random Samples and Independent Observations* and *Independent Samples* are assumed. *Large Sample* (*Normal Distribution*) is given.

Step 3: $t = 0.19$ (–0.19 if samples are reversed.), p-value = 0.573 (for a one sided hypothesis) $\alpha = 0.05$.

Two-Sample T-Test and CI: Salmon, White
```
Two-sample T for Salmon vs White
          N   Mean  StDev  SE Mean
Salmon   15   76.3   27.4      7.1
White    15   74.3   33.3      8.6
Difference = mu (Salmon) - mu (White)
Estimate for difference:  2.1
T-Test of difference = 0 (vs <): T-Value = 0.19  P-Value = 0.573  DF = 27
```

Step 4: Do not reject H₀. There appears to be no difference in reading times for the different colors of paper.

c. Step 1: H0: $\mu_{salmon} = \mu_{white}$, Ha: $\mu_{salmon} < \mu_{white}$, where μ is the population mean reading time.

Step 2: Paired *t*-test: *Random Samples* and *Independence* are assumed. *Large Sample* (*Normal Distribution*) is given. $\alpha = 0.05$.

Step 3: $t = 0.97$ (–0.97 if difference is reversed.), p-value = 0.825 (for a one-sided hypothesis).

Paired T-Test and CI: Salmon, White
```
Paired T for Salmon - White
             N   Mean  StDev  SE Mean
Salmon      15  76.33  27.41     7.08
White       15  74.27  33.27     8.59
Difference  15   2.07   8.27     2.13
T-Test of mean difference = 0 (vs < 0): T-Value = 0.97  P-Value = 0.825
```

Step 4: Do not reject H₀. There appears to be no difference in reading times for the different colors of paper.

d. The paired *t*-test is appropriate because each person is tested twice, so the numbers are coupled.

e. People might get faster if they had read the passage previously, and you don't want the order of reading to affect the answer.

Section 12.2: Controlling Variation in Surveys

12.27 Systematic sampling.

12.29 Those who own an electric car may feel the location of the charging station is more important than students who do not have an electric car and who would be unlikely to use this resource.

12.31 It is systematic sampling.

12.33 It is cluster sampling.

Section 12.3: Reading Research Papers

12.35 a. No, because this is an observational study, not a controlled experiment with random assignment.

b. This association can only be generalized to the entire population of people with Type 2 diabetes if the participants represented a random sample of people with Type 2 diabetes. The sample in this study were all from Japan, so the association cannot be generalized to all people with Type 2 diabetes.

12.37 a. You can generalize to other people admitted to this hospital who would have been assigned a double room because of the random sampling from that group.

b. Yes, you can infer causality because of the random assignment.

12.39 a. The treatment variable is type of intravenous fluid (balanced crystalloids or saline).

b. Yes. 14.3% of the balanced crystalloids group and 15.4% of the saline group had a major adverse kidney event. This difference is significant (p-value = 0.04).

    c. Yes. This controlled experiment used random assignment so a causal conclusion can be made. Use of balanced crysalloids for an intravenous fluid results in a lower rate of major adverse kidney events.

12.41 This is likely the result of an observational study because researchers would not randomly assign subjects to "phub" or "not phub" if the possible outcomes were negative effects on relationship satisfaction and depression. A causal conclusion cannot be made from an observational study.

12.43 This study did not use a control group and random assignment. A causal conclusion cannot be made.

12.45 Randomly assign 100 subjects to participate in YMLI and 100 subjects to do some other type of physical activity for the same amount of time (5 days per week). All study subjects do the activity for 12 weeks. At the end of 12 weeks, measure the biomarkers of cellular aging and longevity in both groups.

12.47 a. The treatment variable was type of music; the response variable was overall divergent thinking (ODT).

    b. This was a controlled experiment. Subjects were randomly assigned to one of the treatment conditions.

    c. The study used random assignment but not random selection. A causal conclusion can be made but we cannot conclude that everyone in general would respond similarly.

12.49 a. Take a nonrandom sample of students and randomly assign some to the reception and some to attend a control group meeting where they do something else (such as learn the history of the college).

    b. Take a random sample of students and offer them the choice of attending the reception or attending a control group meeting where they do something else (such as learn the history of the college).

    c. Take a random sample of students. Then randomly assign some of the students in this sample to the reception and some to the control group meeting.

12.51 a. Is ketamine an effective treatment for Social Anxiety Disorder?

    b. Yes, ketamine may be effective in reducing anxiety.

    c. Patients were given ketamine and placebo infusions in a random order with a 28-day period between infusions. Patients' anxiety was rated 3 hours after the infusion and for a period of 14 days using blinded ratings as well as a self-reported scale.

    d. Yes, the conclusion is appropriate for the study. This was a controlled experiment that used random assignment, so a causal conclusion is appropriate.

    e. Because there was no random sampling from the population of all people with Social Anxiety Disorder, we cannot generalize widely, and the results apply only to these patients.

    f. There was no mention of other articles.

12.53 a. Does self-distancing improve executive function in 3-year old and 5-year old children?

    b. Self-distancing improved executive function in 5-year olds but not in 3-year olds.

    c. This was a controlled experiment. Children were randomly assigned to one of three treatment groups or to a control group. Children in each group did a task and were assessed on their executive function in completing the task by researchers.

    d. The study used random assignment but not random selection. A causal conclusion can be made, but we cannot conclude that all 3-year olds and 5-year olds would respond similarly.

    e. Because there was no random sampling from the population, we cannot generalize widely, and these results apply only to these children.

    f. There was no mention of other articles.

12.55 a. Handover of anesthesia care: 2614/5941 = 44.0%;
No handover of anesthesia care: 89,066/ 307,125 = 29.0%;
The difference in the two groups is statistically significant (p-value < 0.001).

    b. Yes, because the group without complete handover had a lower percentage of bad outcomes.

c. The confidence interval (-0.3% to 2.7%) contains 0 and the p-value is 0.11, leading to a conclusion that there was no significant different in this outcome between the two groups.

12.57 a. The treatment variable was formula type and the response variable was development of Type 1 diabetes by age 11.5 years.

b.

Developed Disease	Conventional Formula	Hydrolyzed Formula	Total
Yes	82	91	173
No	997	1892	2889
Total	1079	1983	3062

Step 1: H_0: Formula type and disease development are independent,
H_a: Formula type and disease development are not independent.

Step 2: Independent, Random assignment, All expected counts greater than 5 (see table).

Step 3: Significance level: 0.05; $X^2 = 11.88$; p-value< 0.001.

Contingency table results:

Rows: Developed Disease
Columns: None

Cell format
Count
(Expected count)

	Conv. Formula	Hydr. Formula	Total
Yes	82 (60.96)	91 (112.04)	173
No	997 (1018.04)	1892 (1870.96)	2889
Total	1079	1983	3062

Chi-Square test:

Statistic	DF	Value	P-value
Chi-square	1	11.881421	0.0006

Step 4: Reject H_0. Formula type and development of Type 1 diabetes are associated.

c. Yes, since infants in this randomized clinical trial who were given hydrolyzed formula were significantly less likely to develop Type 1 diabetes than infants who were given a conventional formula. However, because there was no random sampling from the population, we cannot generalize widely, and these results apply only to these infants.

Chapter 13: Inference without Normality

Answers may vary slightly due to type of technology or rounding.

Section 13.1: Transforming Data

13.1   You should have a random sample from the population.

13.3   The data should be drawn from Normal distributions, or the sample sizes should be large (typically at least 25 from each population). Observations must be independent of each other. The groups must be independent of each other.

13.5   a.   Since it is roughly bell-shaped, Histogram A goes with D, and, since it is right-skewed, Histogram B goes with C.

     b.   A is roughly bell-shaped (Normal), and B is right-skewed.

     c.   Use a log transform on the data that are shown in histogram B.

13.7   a. $\log_{10} 10 = 1$, $\log_{10} 100 = 2$, $\log_{10} 1000 = 3$, $\log_{10} 6500 = 3.8$

     b. $10^3 = 1000$, $10^5 = 100,000$, $10^{2.4} = 251.2$, $10^{3.2} = 1584.9$

13.9   Don't worry if your numbers are a bit different because of rounding.

     a.   (i) $\log_{10} 10 = 1$, $\log_{10} 1000 = 3$, $\log_{10} 10000 = 4$,

         (ii) $\dfrac{1+3+4}{3} = 2.67$

         (iii) Geometric Mean $= 10^{2.67} = 467.7$

     b.   Mean $= \dfrac{10 + 1000 + 10000}{3} = 3670$,   Median $= 1000$; Geometric Mean $= 467.7$,

        from smallest to largest: 467,7 (geometric mean), 1000(median), 3670(mean)

13.11   a. $(791.87, 2541.65)$; I am 95% confident that the population mean amount spent is between $791.87 and $2,541.65.

     b. $\left(10^{2.97}, 10^{3.29}\right) = (933.25, 1949.84)$; I am 95% confident that the population geometric mean is between $933.25 and $1949.84.

     c. $2541.65 - 791.87 = 1749.78$ and $1949.84 - 933.25 = 1016.59$; The confidence interval for the geometric mean (from the logarithms) is narrower.

     d.   The confidence interval for the geometric mean is more appropriate, because the distribution of the log-transformed data is more symmetric, and so the confidence level for the geometric mean will be more accurate. Also, we can get a more precise (smaller margin of error) estimate for the geometric mean than for the mean, based on this sample.

13.13   a.   Right-skewed

     b. $(4.04, 6.31)$

One-Sample T: Hour

Variable	N	Mean	StDev	SE Mean	95% CI
Hour	45	5.178	3.774	0.563	(4.044, 6.312)

     c.   It is closer to Normal than the distribution of untransformed data.

d. (0.515, 0.7025)

```
One-Sample T: Loghour
Variable    N    Mean   StDev   SE Mean      95% CI
Loghour    45   0.6087  0.3121   0.0465   (0.5150, 0.7025)
```

e. $\left(10^{0.515}, 10^{0.7025}\right) = (3.27, 5.04)$; We are 95% confident that the population geometric mean number of hours of exercise per week for all students at this college is between 3.27 hours and 5.04 hours.

f. Answers may vary. The interval for the geometric mean is more precise (smaller margin of error), which suggests that the geometric mean might be a better measure of center. On the other hand, the sample size is large, so both confidence intervals are valid and the population mean could also be used. (If the sample size were under 25, the reported confidence level for the resulting confidence interval of the population mean would not be accurate.)

Section 13.2: The Sign Test for Paired Data

13.15   Step 1: H_0: The median levels of lead in the blood are the same.
        H_a: The median level of lead in the blood is larger for the experimental group (one-sided).

Step 2: Sign test: *Random Sample* and *Paired Data* are met and *Independent Measurements* is assumed. $\alpha = 0.05$

Step 3: $S = 28$ or 4, p-value < 0.001.
For a p-value from a TI-84, use binomcdf(32, 0.5, 4) = 9.7×10^{-6}, or about 0.00001.

```
Sign Test for Median: Exposed - Control
Sign test of median =  0.00000 versus > 0.00000
                    N  Below  Equal  Above      P   Median
Exposed - Control  33    4      1     28    0.0000   15.00
```

Step 4: Reject H_0. The median level of lead in the blood was larger for the children in the experimental group.

13.17   a. The medians were 19 (for same) and 33.50 (for different). Yes, it tended to take longer for the different colors.

b. Step 1: H_0: Median time for same equals median time for different.
    H_a: Median time for same is less than median time for different.

Step 2: Sign test: *Random Sample* is assumed, *Paired Data* are met, and *Independent Measurements* is assumed. $\alpha = 0.05$

Step 3: $S = 1$ or 9, p-value = 0.0107.
For a p-value from a TI-84, use binomcdf(10, 0.5, 1) = 0.0107

```
Sign Test for Median: Same - Diff
Sign test of median =  0.00000 versus < 0.00000
               N  Below  Equal  Above      P   Median
Same - Diff   10    9      0      1    0.0107  -11.50
```

Step 4: Reject H_0. The difference is significant. It took longer to identify the wrong color.

13.19   Step 1: H_0: The median pulse rate does not change. H_a: The median pulse rate goes up (one-sided).

Step 2: Sign test: *Random Sample* is assumed, *Paired Data* are met, and *Independent Measurements* is assumed. $\alpha = 0.05$

Step 3: $S = 4$ or 5, p-value = 0.500.
For a p-value from a TI-84, use binomcdf(9, 0.5, 4) = 0.5000

Sign Test for Median: AfterMen - BeforeMen

```
Sign test of median =  0.00000 versus > 0.00000
                     N  Below  Equal  Above       P  Median
AfterMen - BeforeMen 9      4      0      5  0.5000   4.000
```

Step 4: Do not reject H_0. The median pulse rate did not go up significantly for men.

13.21 Step 1: H_0: Median age for grooms = median age for brides.
       H_a: Median age for grooms > median age for brides.

Step 2: Sign test: *Random Sample* is assumed, *Paired Data* are met, and *Independent Measurements* is assumed. $\alpha = 0.05$

Step 3: $S = 10$ or 3, p-value = 0.046, from a TI-84 using binomcdf(13, 0.5, 3).

Step 4: Reject H_0. The median age for the grooms is significantly greater than the median age for the brides.

Section 13.3: Mann-Whitney Test for Two Independent Groups

13.23 a. The median for these ethicists was 4.5 meals per week, which is lower than the median for this control group (6.5). In the sample, ethicists reported eating fewer meals with meat per week than nonethicists reported.

       b. Step 1: H_0: Median meals of meat per week of control = median of ethicists.
           H_a: Median meals of meat per week of control > median of ethicists.

       Step 2: Mann-Whitney test: *Random Samples, Independent Observations, Independent Groups, Numerical and Continuous*, and *Same Shape* are met. $\alpha = 0.05$.

       Step 3: $W = 198$, p-value = 0.058.

       Step 4: Do not reject H_0. We have not found a significant difference in behavior.

13.25 a. This shape is right-skewed. It appears there may be two outliers at about 17 runs. Because of the skew and outliers, it would be more appropriate to compare medians.

       b. Step 1: H_0: The median for winning runs is the same for both the American and National leagues.
           H_a: The median for winning runs is different for the two leagues.

       Step 2: Mann-Whitney test: *Random Samples, Independent Observations, Independent Groups, Numerical and Continuous*, and *Same Shape* are met.

       Step 3: $\alpha = 0.05$, $W = 216$, p-value = 0.552.

       Step 4: Do not reject H_0. There is not sufficient evidence to conclude the medians are significantly different.

13.27 a. We don't compare means because there is an outlier of around $400 for the women, and the samples are not large.

       b. Step 1: H_0: Median for all men = median for all women. H_a: Median for all men < median for all women.

       Step 2: Mann-Whitney test: *Random Samples, Independent Observations, Independent Groups, Numerical and Continuous*, and *Same Shape* are met. $\alpha = 0.05$..

       Step 3: $W = 501.0$, p-value = 0.141.

       Step 4: Do not reject H_0. The medians for men and women with regard to cell phone bills are not significantly different.

13.29 a. The histogram is strongly left-skewed, so it would be more appropriate to compare medians.

       b. The sample median of 80 for the women is higher than the sample median of 75 for the men.

       c. Step 1: H_0: The median for all men is equal to the median for all women.
           H_a: The median for all men is not equal to the median for all women.

Step 2: Mann-Whitney test: *Random Samples, Independent Observations, Independent Groups, Numerical and Continuous*, and *Same Shape* are met.

Step 3: $\alpha = 0.05$, $W = 106,108$, p-value $= 0.032$.

Step 4: Reject H_0. The medians are significantly different.

Section 13.4: Randomization Tests

13.31 a. The red line looks like it is pretty far out in the tail of the data and suggests visually that there is a real difference in extraversion between the sporty and the nonsporty students.

   b. The p-value is 0.011988012, or 0.012.

   c. H_0: $\mu_{\text{sports yes}} = \mu_{\text{sports no}}$, H_a: $\mu_{\text{sports yes}} > \mu_{\text{sports no}}$, where μ is the population mean extravert level. We reject the null hypothesis and conclude that sporty students typically have a higher level of extraversion than nonsporty students.

   d. We could get an approximate p-value using the histogram, by finding the (approximate) proportion of observations to the right of the red vertical line.

13.33 a. The distributions are strongly right-skewed, and the median is often a better choice for skewed data than the mean.

   b. 525 is not far out in the tail, so it is not unusually large.

   c. H_0: Median credit card debt for men equals median credit card debt for women. H_a: Median credit card debt for men does not equal median credit card debt for women. The one-tailed p-value is larger than 0.05, so the two-tailed p-value will be even larger. Do not reject H_0. The population median for the men has not been shown to be different from the population median for the women.

13.35 H_0: $\mu_{\text{seeded}} = \mu_{\text{unseed}}$, H_a: $\mu_{\text{seeded}} > \mu_{\text{unseed}}$, where μ is the population mean rainfall. Choose 0.025 by looking at the graph, knowing that the observed number is 368.9 acre-feet. Reject the null hypothesis. The mean rainfall is greater for the seeded clouds.

13.37 Explanation a.

Chapter Review Exercises

13.39 a. Paired *t*-test or sign test

   b. Sign test

   c. Paired *t*-test or sign test

13.41 a. Two-sample *t*-test or Mann-Whitney test

   b. Mann-Whitney test

13.43 a. Two-sample *t*-test or Mann-Whitney test

   b. Mann-Whitney test

13.45 a. The sign test

   b. The paired *t*-test or the sign test

   c. The paired *t*-test or the sign test. Even though the differences are skewed, the sample size is large enough to use the *t*-test.

13.47 The answers follow the guidance on page 654.

Step 0:

Obs	Null Value	Difference
4.2	3.18	$4.2 - 3.18 = 1.02$
3.6	3.18	$3.6 - 3.18 = 0.42$
3.9	3.18	$3.9 - 3.18 = 0.72$
3.4	3.18	$3.4 - 3.18 = 0.22$
3.3	3.18	$3.3 - 3.18 = 0.12$

Step 1: H_0: The median weight of the cones is 3.18 ounces, as advertised (or the median difference is 0).
H_a: The median weight is more than 3.18 ounces (or the median difference is more than 0).

Step 2: Use the one-sample sign test. The cones are independent because they are purchased from different servers, and although we don't have a truly random sample, we hope we have a representative sample.

Step 3: $\alpha = 0.05$, $S = 0$ or 5, p-value $= 0.031$

For a p-value from a TI-84, use binomcdf(5, 0.5, 0) = 0.0313

```
Sign Test for Median: Cone Weight
Sign test of median =   3.180 versus > 3.180
              N   Below  Equal  Above      P  Median
Cone Weight   5     0      0      5   0.0313   3.600
```

Step 4: Reject H_0 and conclude that the population median is more than 3.18 ounces (or the population median difference is more than 0.00).

Why use the sign test? We have a small sample and do not know whether the distribution is Normal or not.

13.49 a. They are both right-skewed.

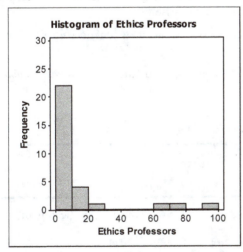

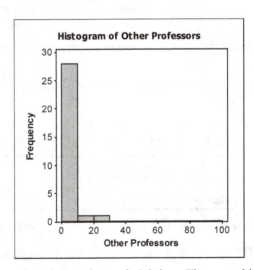

b. The mean for the ethicists is 12.7 days, and the mean for other professors is 3.2 days. The non-ethics professors tend to have been in more recent contact with their mothers.

```
Descriptive Statistics: Ethics Professors, Other Professors
Variable             N   Mean   Median
Ethics Professors   30  12.73    5.00
Other Professors    30   3.200   1.000
```

c. The median for the ethicists is 5 days, and the median for other professors is 1 day, so other professors tend to have been in more recent contact with their mothers.

d. Step 1: H_0: $\mu_{eth} = \mu_{other}$, H_a: $\mu_{eth} \neq \mu_{other}$

Step 2: Two-sample *t*-test: *Random Samples and Independent Observations*, *Independent Samples* and *Large Sample* (*Normal Distribution*) are assumed.

Step 3: $t = 2.23$ or -2.23, p-value $= 0.033$.

```
Two-Sample T-Test and CI: Ethics Professors, Other Professors
Two-sample T for Ethics Professors vs Other Professors
                        N    Mean   StDev   SE Mean
Ethics Professors      30    12.7    23.0     4.2
Other Professors       30    3.20    4.34     0.79
Difference = mu (Ethics Professors) - mu (Other Professors)
Estimate for difference:   9.53
95% CI for difference:   (0.81, 18.25)
T-Test of difference = 0 (vs not =): T-Value = 2.23   P-Value = 0.033   DF = 31
```

Step 4: Reject H_0. There is a significant difference in means.

e. Step 1: H_0: $Med_{ethics\ professors} = Med_{other\ professors}$, H_a: $Med_{ethics\ professors} \neq Med_{other\ professors}$

Step 2: Mann-Whitney test: *Random Samples, Independent Observations, Independent Groups, Numerical and Continuous*, and *Same Shape* are assumed.

Step 3: $\alpha = 0.05$, $W = 1126.5$, p-value $= 0.002$.

```
Mann-Whitney Test and CI: Ethics Professors, Other Professors
                       N    Median
Ethics Professors     30     5.00
Other Professors      30     1.00
Point estimate for ETA1-ETA2 is 2.00
95.2 Percent CI for ETA1-ETA2 is (0.99,4.00)
W = 1126.5
Test of ETA1 = ETA2 vs ETA1 not = ETA2 is significant at 0.0018
The test is significant at 0.0016 (adjusted for ties)
```

Step 4: Reject H_0. There is a significant difference in medians. The typical time since contact with mothers is different for ethics professors than for other professors.

13.51 Step 1: H_0: Median texts sent equals median texts received, H_a: Median texts sent does not equal median texts received

Step 2: Sign test: *Random Sample* is assumed, *Paired Data* are met, and *Independent Measurements* is assumed.

Step 3: $\alpha = 0.05$, $S = 40$ or 11, p-value < 0.001.
For a p-value from a TI-84, use binomcdf$(11, 0.5, 4) = 0.00002$.

```
Sign Test for Median: Sent - Received
Sign test of median =  0.00000 versus not = 0.00000
                    N   Below   Equal   Above      P    Median
Sent - Received   109    40      58      11    0.0001   0.00000
```

Step 4: Reject H_0. The median number of texts sent and the number received are significantly different.

13.53 Step 1: H_0: $\mu_{sent} = \mu_{received}$, for males, H_a: $\mu_{sent} \neq \mu_{received}$ for males.

Step 2: Paired *t*-test (matched pairs), *Random Samples and Independent Observations*, *Independent Samples*, and *Large Sample* (*Normal Distribution*) assumed.

Step 3: $\alpha = 0.05$, $t = 1.20$ or -1.20, p-value $= 0.234$.

```
Paired T-Test and CI: Send_Male, Receive_Male
Paired T for Send_Male - Receive_Male
                   N     Mean    StDev   SE Mean
Send_Male         61    27.70    68.32    8.75
Receive_Male      61    28.66    67.94    8.70
Difference        61   -0.959    6.236    0.798
95% CI for mean difference: (-2.556, 0.638)
T-Test of mean difference = 0 (vs not = 0): T-Value = -1.20   P-Value = 0.234
```

Step 4: Do not reject H_0. We do not have enough evidence to conclude that the means are significantly different.

13.55 You cannot find the geometric mean because you cannot find a logarithm of 0. And there are several students who missed no classes.

13.57 The red line is far out in the left tail, which suggests a significant difference. The p-value for the one-sided hypothesis is 0.00899. The conclusion is that ethicists tend to have less recent contact with their mothers than other professors. (The mean number of days since contact is greater for the ethicists than for the non-ethicists.)

13.59 a. There was a significant difference in purchase intention toward V Energy drink. The p-value was less than 0.05 indicating there was a difference between the control group and experimental group for this item.

   b. To use the independent sample t-test, the population distributions must be Normal or the sample sizes must be large enough for the Central Limit Theorem to be applied. The Mann-Whitney test only requires that the two population distributions have the same shape, not necessarily a Normal distribution.

Chapter 14: Inference for Regression

Answers may vary slightly due to type of technology or rounding.

Section 14.1: The Linear Regression Model

14.1 Answers will vary. Some possible random factors that might affect test scores are: the amount of time the student could study, the amount of sleep the student got the night before, the particular choice of questions on the exam, the noise level in the room, and the health of the student.

14.3 Answers will vary. Some possible random factors that might affect height: diet, genetics, and medications.

14.5

Floors	Height (in feet)	Predicted Height	Residual
163	2717	2690.8	26.2
101	1398	1454.9	−56.9
101	1358	1454.9	−96.9
88	1289	1195.8	93.3
89	1250	1215.7	34.3

14.7 a. The residual plot shows a curvature, indicating the linear condition fails. The linear model is not appropriate.

b. There are two points that look farthest from the regression line: a car that is about 7 years old and a car that is about 15 years old.

14.9 The residual plot is fan-shaped, showing more variation in the number of units for students who have attended for many semesters, and less variation for those who have attended for few semesters. This shows that the constant standard deviation condition does not hold, so inference would not be appropriate.

14.11 Linear regression is not appropriate, because the constant standard deviation condition does not hold; there is more variation with the larger beginning salaries than with the smaller beginning salaries.

14.13 The residual plot shows an increasing trend, and the QQ plot does not follow a straight line. Linear regression is inappropriate for this data set, because the linearity condition and the Normality condition are not met.

Section 14.2: Using the Linear Model

14.15 Step 1: H_0: $\beta_1 = 0$, H_a: $\beta_1 \neq 0$

Step 2: t-test for slope: *Linearity, Constant Standard Deviation, Normality,* and *Independence* are given.

Step 3: $\alpha = 0.05$, $t = 13.6$, p-value<0.0001.

Step 4: Reject H_0. We have enough evidence to reject the hypothesis that the slope is 0. We have shown a linear association between state and federal spending on education.

14.17 a. The intercept is 2.78. If the intercept were 0, it would mean that if a mother had 0 years of education, then the father would be predicted also to have 0 years of education.

b. Step 1: H_0: $\beta_0 = 0$, H_a: $\beta_0 \neq 0$

Step 2: t-test for intercept: *Linearity, Constant Standard Deviation, Normality,* and *Independence* are given.

Step 3: $\alpha = 0.05$, $t = 2.00$, p-value $= 0.056$.

Step 4: We cannot reject H_0. We don't have enough evidence to reject an intercept of 0.

14.19 a. The slope is 0.936. On the average, for each inch taller a parent is, the child is about 0.94 inch taller, in the sample.

b. Step 1: H₀: $\beta_1 = 0$, Hₐ: $\beta_1 \neq 0$

Step 2: *t*-test for slope: *Linearity*, *Constant Standard Deviation*, *Normality*, and *Independence* are given.

Step 3: $\alpha = 0.05$, $t = 8.38$, p-value < 0.001.

Step 4: Reject H₀. There is a significant linear relationship between parent height and student height.

c. If the slope were 1, it would mean that on the average, for every inch taller the parent was, the student would be 1 inch taller also.

d. Step 1: H₀: $\beta_1 = 1$, Hₐ: $\beta_1 \neq 1$

Step 2: *t*-test for slope: *Linearity*, *Constant Standard Deviation*, *Normality*, and *Independence* are given.
$\alpha = 0.05$

Step 3: $t = \dfrac{\hat{\beta}_1 - \text{null}}{SE_{\hat{\beta}_1}} = \dfrac{0.9355 - 1}{0.1116} = -0.578$, p-value > 0.05

Cumulative Distribution Function
```
Student's t distribution with 27 DF
     x    P( X <= x )
-0.578      0.284027
```

The p-value is found from the given table because 0.578 is less than 2.052.

Step 4: Do not reject H₀.

14.21 a. Slope = 0.22. The slope tells us that for each additional library visitor, the number of wifi sessions increase on average by about 0.22. In other words, for 10 additional library visitors, the number of wifi session increase on average by about 2.

b. (0.03, 0.41)

c. The interval does not contain 0 so there is a relationship between number of visitors and number of wifi sessions.

14.23 Yes. The output shows that we cannot reject the null hypothesis that the intercept is 0 (because the p-value is 0.740, which is larger than 0.05). This means that the confidence interval will include 0 pounds of trash.

Section 14.3: Predicting Values and Estimating Means

14.25 Use a prediction interval. This concerns prediction of a single value, not a mean.

14.27 She should use a prediction interval because she is predicting one value, not a mean.

14.29 Use a confidence interval. This concerns prediction of a mean, not a single value.

14.31 a. Price $= 16,209 + 101.9(2500) = \$270,959$

b. Prediction interval, because we are predicting the value for one house, not the mean of a group of houses.

c. ($160,581,$381,532); *Linearity*, *Constant Standard Deviation*, *Normality*, and *Independence* are assumed.

d. Yes, he can afford a house because he has access to enough money to pay the price at the top of the 95% interval.

14.33 The prediction interval is about 125 to about 200, or (125, 200).

14.35 The prediction interval is about 1.8 to about 3.8, or (1.8, 3.8). This is a very wide interval, and it is not very useful to find that a student with a 750 on the math SAT will have a GPA between a D and an A.

14.37 The green lines are prediction intervals (for individuals), and the red lines are confidence intervals (for means). Means tend to be more stable than individual measurements and give more precise results, which is why their intervals are narrower.

14.39 a. The prediction interval is $(182, 216)$; the confidence interval is $(195, 203)$. The confidence interval is narrower because it is estimating a population mean, and there is less uncertainty in this than in the prediction interval, which is predicting the weight of an individual.

   b. We are 95% confident that the mean weight of all baseball players who are 20 years old is between 195 pounds and 203 pounds.

   c. We are 95% confident that one 20-year-old baseball player will weigh between 182 and 216 pounds.

Chapter Review Exercises

14.41 a. See graph. The equation is: Longevity $= 7.84 + 0.0327$ Gestation

   b. The hippopotamus is predicted to live about 10 years but lives over 40 years, on average.

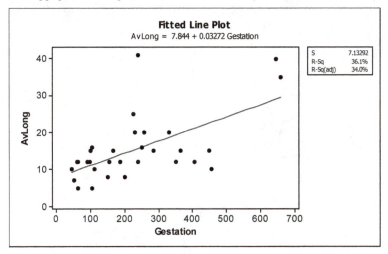

   c. The confidence interval is 13.92 to 19.18 years, or $(13.92, 19.18)$.

```
Regression Analysis: AvLong versus Gestation
The regression equation is
AvLong = 7.84 + 0.0327 Gestation
Predictor      Coef    SE Coef     T      P
Constant      7.844     2.233    3.51   0.001
Gestation  0.032724   0.007943   4.12   0.000
S = 7.13292    R-Sq = 36.1%    R-Sq(adj) = 34.0%


Analysis of Variance
Source          DF      SS       MS      F      P
Regression       1   863.52   863.52  16.97  0.000
Residual Error  30  1526.36    50.88
Total           31  2389.88


Predicted Values for New Observations
New
Obs    Fit   SE Fit      95% CI            95% PI
  1  16.55    1.29   (13.92, 19.18)    (1.75, 31.35)
```

   d. Humans do not fit into this pattern and are not included in the data set. The average human life expectancy is much more than the top of the confidence interval, which is 19.18 years.